25 Years: A Lifetime At Sea

Best Wishes

by

Kenneth Pickering

K. Pickering

Published by

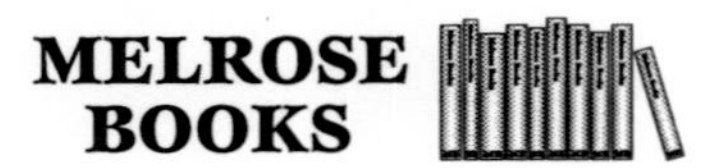

An Imprint of Melrose Press Limited
St Thomas Place, Ely
Cambridgeshire
CB7 4GG, UK
www.melrosebooks.co.uk

FIRST EDITION

Cover designed by David Pearce

ISBN 978-1-908645-47-0

Printed and bound in Great Britain by:
TJ International Ltd, Padstow, Cornwall

FSC www.fsc.org MIX Paper from responsible sources FSC® C013056

PREFACE

Some people ask, what's the earliest you can remember? The earliest I can remember is when I was about three years old. My mother was bathing me in a tin bath in front of the coal fire when one of her friends, Mrs Torr, popped round to visit. Like both my parents, sadly the lady has since passed away.

I was born at No. 6 Close Street, Hemsworth in 1943 and lived at 4 St. Helen's Avenue for approximately 66 years. It was one of the first council housing estates to be built in Hemsworth under the Housing Act, just after the 1st World War.

I was born in 1943, during the 2nd World War. I don't remember much about my early upbringing with my four brothers and three sisters.

The first school I attended was R.H. Gilbert – opposite the fire station – which closed and was used in later years until the present day as a youth club. I then went on to attend Archbishop Holgate School at the end of Cemetery Road. Whilst attending Archbishop Holgate School, the school organised a trip across to Belgium. Those who went were fortunate that their parents paid for the week's visit to Bruges, as money was tight at that time in the 1950s.

For my last four years at school, I attended West End Secondary Modern, at the top of Regent Street.

It was ideal for me as I only had to climb over the top end of my garden wall and I was in the school playground.

Unfortunately, this school was pulled down in the early 2000s, but a brand new school was built in its place for infants and juniors.

Archbishop Holgate School was also demolished, but wasn't replaced.

When I left school at 15 years of age with no academic qualifications, I hadn't a clue what I was going to do. My father was adamant I wasn't going to work down the coal mines, as my three elder brothers had.

By some luck – or bad luck – I managed to get a job in a local factory, named Samuel Hunt's, opposite the Catholic Church.

This local firm's products were tripe and sausage casing for sausages, some of which were exported across to the continent. The same factory is now a carpet store.

After working at the factory for two years, Bernard Robbin, a friend who was a year younger than me, said he was joining the Merchant Navy.

I got as much information as I could from him, put in for a day off work, and went through to Goole to the Shipping Federation, where I filled all the forms in without telling my parents or the firm I was working for. After I was accepted, my parents weren't too keen for me to go, but in the end they relented. So I put my notice in, and went down to Sharpness in Gloucestershire to prepare for six weeks' Pre-Sea Training on the Vindicatrix, from 6th Febuary, 1961 to 17th March, 1961.

CHAPTER 1

The Vindicatrix was a three-masted sailing ship and was about 74 years old when taken out of dock and to the brokers at Newport in Wales in January 1967.

During her service, the Vindicatrix was based at Sharpness in Gloucestershire on the River Severn. It was used from around 1939 as a Training School, preparing 70,000 seamen, mainly for the Merchant Navy.

On leaving the school, some trainees just over 15 years old were losing their lives during the 2nd World War in the Merchant Navy.

At Sharpness, a plaque was unveiled to commemorate the service of the Vindicatrix Boys for their service to the 2nd World War, and to all those who lost their lives at sea during the conflict.

Approximately 500 boys trained there each year at the National Sea Training School, and were known as the "Vindi Boys". Some trained as deck boys, and the remainder were catering boys. The deck boys did 10 weeks' Pre-Sea Training whilst the catering boys did six weeks.

After training you would go into the Merchant Navy and be appointed to join a ship, after you passed out.

The camp where the training commenced was, I think, an old Army Camp, where the accommodation for the trainees was billeted.

We were told not long after joining that if you could stick the training, you could stick anything.

Morning call was 0530, and out on parade at six o'clock in all weathers. I was lucky to get into the school, as at the time I had just turned 17 years old and 17½ was the cut-off.

I was there from February 1961 (6.2.61 to 17.3.61) to early March, prior to joining my first voyage on 23rd March, 1961.

There wasn't really much to do down at Sharpness. Sometimes you were allowed into the town at weekends to see a film, but you stuck out like a sore thumb in your thick navy blue serge uniform; you weren't allowed out of camp in civvies.

Every Sunday morning was Inspection of Billets and out on parade.

Once a month 100 tons of coal was shipped and loaded onto the Vindicatrix to fuel the boilers and galley.

Not long after arriving at the camp, you would be sent to see the barber, named Charlie, for a neat trim. There was always a crowd egging him on to cut it short. Some of the trainees who had long hair decided that was enough for them, and left for home.

Depending when you attended the Training School, you might get the chance to enjoy the annual Sports Day, which was held every June/ July and was one of the camp's highlights of the year.

One of the other highlights down at Sharpness was if you received mail or a food parcel from home, as things were a bit tight in those days. After being in the camp for about four weeks of training, you would actually start living on the Vindicatrix, to get used to life at sea. The deck boys doing their training would be involved in splicing of ropes, or making different types of knots, rigging and staging of Bosun's Chairs when doing painting at sea. They also had to learn about winches, relating to tying up ships when entering ports around the world.

After their 10 weeks' training and passing out, they would be appointed to a ship as a deck boy, which would be for a period of around nine months, possibly on two or three different types of ships to gain experience.

After that nine months the deck boy would sit a test and, if passed, he would be promoted to J.D.S. or Junior Ordinary Seaman. This would be followed by further sea time and he would then be promoted to S.O.S. Senior Ordinary Seaman. If he passed his next exam, he would be promoted to E.D.H. Efficient Deck Hand. Some unfortunately failed and had then to go to sea as D.H.U. Deck Hand Uncertified until they passed their exam. The last promotion was to A.B. as Able Bodied Seaman. Some went on to reach Bosun, or in later years were named C.P.O. Chief Petty Officer.

During their Pre-Sea Training, the catering boys learned different ways of laying tables and placings, as they could be sent to different types of ships, including oil tankers, cargo or passenger liners. They were also taught how to clean cabins, make different types of bunk/beds, clean toilets and alleyways, and waiting on tables.

On joining a ship, they would be shown their daily duties for the days at sea. Their next promotion would be to an Assistant Steward normally, which could take up to 18 months to two years on maybe five or six different types of ships.

On reaching Asst/Steward, the next promotion would be to 2nd Steward, although different companies would have alternative ways of promotion on their ships. The 2nd Steward would have his own duties, which included looking after the accommodation for the Captain and Chief Engineer.

In some shipping companies, promotion could be quite swift if you signed a contract with them. In the mid-1960s to the 1970s, they all seemed to have their own stages to suit their ships.

The next promotion after 2nd/Steward would be 2nd/Cook and Baker, which would involve training ashore in a catering college. This would normally take 10 to 12 weeks' training and, after an examination was taken and passed, then it would be back at sea from 12 months to 2 years on different types of ships.

Next promotion would be to attain a Ship's Cook's Certificate, which would also involve training ashore – normally a six week course.

After gaining the certificate, it would be back at sea. To show how important this certificate was, merchant vessels wouldn't be able to leave port in the UK without a certificated Ship's Cook on board.

Many Ship's Cooks went on to become Chief Stewards, a title which in later years was changed to Catering Officer.

Although the Ship's Cook's Certificate was important, as a qualification it wasn't much good ashore in the catering industry. Ashore, the main establishments needed their staff to have the City & Guilds Certificate of the London Institute.

Sadly, the Sea Training School at Sharpness closed in 1966, and a new school opened at Gravesend in Kent.

CHAPTER 2

On leaving the National Sea Training School in March 1961, I was issued with a Seaman's Record/Discharge Book, an Identity Book, and an International Certificate of Vaccination, which had to be kept up-to-date for Smallpox, Cholera and Yellow Fever.

After being at home for a few days, you would be notified by the Shipping Federation to join a ship.

The nearest place for me was Goole, which was where the Shipping Federation was established. The first ship I was to join was a Shell tanker, STS *Achatina*, which was down at Tilbury Docks. All of the ships in the Shell Tanker Company were named after sea-shells.

After going on board, you would have to officially sign on, and my first voyage – entered in the Seaman's Record Book – started on the 22.3.61. My first salary at sea would start at £14 per month, for a 56-hour week. At the time I was sending £6 per month home to my mother. It may not have sounded a lot of money at the time, but there was also food and lodgings to pay, which weren't included in that £14 a month. The hours of work per day were eight hours, seven days a week.

When you signed on, the description of the voyage would be stated as foreign-going, home trade, or a run job.

After leaving Tilbury Docks, the ship sailed round to Ellesmere Port in Liverpool to load a cargo, and then to a port in the Dutch West Indies, named Curacao.

The full complement of Officers and Crew Members was approximately 45, but the difference in the total men on board could vary from the size of tonnage of different ships.

The accommodation could also vary; STS *Achatina* had midship living quarters for Officers, and aft living quarters for Crew.

There were two Catering Boys on the *Achatina* – myself and what was named a Galley Boy. Basically we did the same jobs but they varied, depending on what duties you performed. We shared accommodation with bunk beds, one up one down, wardrobes and a desk with drawers. It was basic but adequate.

Life at sea sometimes could become monotonous, with the same duties day in, day out. In my case it started at six o'clock in the morning, scrubbing alleyways up the midships, then down aft in the saloon/pantry, washing crockery and cutlery from the breakfast servery in the Officers' Dining Saloon.

The Officers would be served at the table by an Assistant Steward. The 2nd/Steward would plate the food up and prepare the vegetable dishes, and the Assistant Steward would serve up at the table.

In the Crew's Mess Room, the crew would queue up and the Assistant Steward would plate the food as in a canteen style.

There would normally be two sittings at meal times for the different watches at sea – normal hours for watch keepers, 12-4 hrs, 4-8hrs and 8-12hrs, twice a day.

Going across the Atlantic Ocean for the first time was quite daunting, as we hit some bad weather. The first time I was hit by sea-sickness, I never thought I would get over it. I was bad for about three days, but there wasn't much you could do about it and you still had to work. Then about halfway across, the bad weather abated and it was plain sailing in glorious sunshine.

Recreation time at sea was a bit limited. Apart from a film once a week, you could go to the Recreation Room, where there were a darts board, dominoes, chess, draughts or card games.

Every week there would be a bond issue. This consisted of cigarettes, or hand-rolling tobacco, on a set day of the week. If you were over 18 years old, you would also be allowed three cans of beer which would be issued on another day of the week.

Some of the Seamen would save their beer up and have a blow-out or, as it is termed, 'went on the piss'. Some would be so bad they wouldn't be able to turn up for work, so they would be logged and fined two days' pay with a day's leave forfeited.

Cigarettes used to cost 10 shillings for a carton. To buy the same today, it would cost about £50.

We reached the island of Curacao, which was part of the Dutch West Indies, after 10 days at sea. Now commonly known as the Netherlands Antilles – the capital being named Willemstad – a large part was owned by the Royal Dutch Shell Company, who had a huge oil refinery on the island. Ships would bring crude oil into the refinery, where it would then be refined into different types of fuels. In the refinery a section was laid off for local staff where they could swim in the sea; it was fenced off in case sharks were about in the sea. The seamen could use these facilities in their time off for recreation, and I vividly remember using them. The weather out there was out of this world, and you could relax and enjoy being away from shipboard life.

You would normally be in port for about 36 to 48 hours, but on this occasion we discharged the cargo and then back loaded to take a cargo of petrol products up the River Amazon in Brazil.

Once the ship left Curacao, we were only a few days before reaching the mouth of the River Amazon. It was so that wide you could hardly see the land from both ends.

We had to go 1,000 miles up the Amazon River, to a port called Manus. At that time any further up the river was out of the Brazilian authority's jurisdiction, as head-hunters were still about.

On reaching the mouth of the Amazon, two pilots boarded the ship to take the vessel up the river. Both working 12 hours on, 12 hours off for the three day journey.

It was something that will last forever in my memory; going up the river, seeing the different types of moths, butterflies, insects, beetles, etc, in the early mornings.

On arriving at Manus, I personally didn't venture ashore, as it was a bit primitive in the sixties.

On the way back down, the company instructed the Master to load a cargo of fresh water at the mouth of the river and take it back to Curacao, as there was a shortage of water on the island.

Whilst we were at anchor loading the water, it was decided to have a film. If the weather was warm we would show the film outside, but on this occasion we were warned by the local authorities to be wary of a certain kind of fly, which was about the size of a bluebottle. Apparently it could bite the skin and lay an egg underneath, causing the affected area to swell up and require treatment. Fortunately nobody was affected.

On the ship *Achatina*, two of the catering staff were gay. As this was the first time I had encountered this, I hadn't a clue what to expect until I got a rude awakening. I then understood what homosexual meant.

They could be quite comical at times and some of the crew engaged in sexual pleasure with them, but I am afraid it wasn't my cup of tea.

Curacao was the main port in the West Indies for loading and discharging cargo, and when we got back there we were sent once again to Brazil, this time to a port named Recife.

One day while we were at sea, the Chief Steward came to me and said he had a special job for me to do. I had to go down to the engine room and see the engineer on watch for "a long stand". After I had been down there for about 30 minutes, I still didn't get it, until the engineer told me I'd had my long stand and could go.

Sundays at sea were when the Captain, Chief Engineer, Chief Officer and Chief Steward started their Inspection of Officers' and Crews' Quarters. I was a bit worried on some occasions as this could be quite daunting; everything had to be spick and span – *bain maries*, hot-presses, cupboards and lockers, all clean and tidy. And the Captain wore white gloves, so everything had to be clean!

After being on the *Achatina* for just over three months, the ship received the orders to load another cargo at Curacao, then head back across the Atlantic to discharge a cargo at Ellesmere Port near Liverpool.

The length of the voyage on the STS *Achatina* was four months to the date of signing on to signing off – 22/3/61 to 22/7/61.

On signing off the ship, you would be paid in cash, less all the deductions, remittance, bond account, W.L. stamps, etc.

You would then be given a travel warrant to the nearest main line station, to head home on leave.

CHAPTER 3

After being at home for just over six weeks, I was notified to join one of BP's ships up in North Shields, on the Tyne. It was called *British Bulldog*. The ship at the time was in dry dock, in Smiths Dock, having repairs prior to going back to sea. We had just left the Tyne Pier when the ship broke down. So the Engineering Staff had to sort out the problem before we set sail for the Persian Gulf.

To reach the Gulf to load a cargo, it would take approximately three weeks via the Suez Canal. The crew onboard was predominantly from the North East, mainly from the Newcastle area. They were a good bunch of lads and down to earth, but it took me a while to understand the Geordie dialect and accent.

Transit through the Suez Canal was an experience. The first time it appeared to be chaos, as there were so many Egyptians on board on arrival there – the Port Authorities, Customs, even bum boats selling all their souvenirs.

Depending which way you transited the Suez Canal, Southbound, you would travel for so long then you would tie up in what was known as the Bitter Lakes. Then a convoy would pass Northbound and go straight through, with as many as 20 to 30 ships. These convoys varied in size.

After the Northbound convoy passed, the Southbound ships could progress down into the Red Sea. After passing through the Red Sea, it was round to the Persian Gulf to load the cargo. At that time, in the 50s/60s, BP had a big influence in Persia with the Anglo/Iranian Oil Company.

You would normally load a cargo at Bandar Mahshahr, as this was a Crude Oil Port. Further up the Shatal Arab River was a products refinery named Abadan, part of the Anglo/Iranian Oil Company. Once the cargo was loaded we had to proceed down to the Port of Kwinana, near Fremantle, in Western Australia; this part of the voyage would take about 10 days.

About every two or three days the clocks would go on one hour for when you arrived in Australia. That way you would be on the same time zone as them, and vice versa when you went back to the UK.

On arriving in Australia, we found a hive of activity and there was the possibility of getting ashore for the first time there. Mail from home played a big factor, as sometimes it could be five to six weeks without any word from home if mail was missed at the last port of call.

Ship stores had to be loaded on. As Australia was a good place to stock up, the ship was to store for 26 weeks. Unfortunately, it all had to be carried up the gangway as, when the ship started discharging the cargo, the cranes weren't allowed to be used as a safety factor. So it was an all day job, mainly done by the Catering Department because, apart from the Deck Hands on watch, the rest of the crew had been given time off.

The stores had to be put away in the Dry Storeroom, frozen meat into the deep fridges, flour into the ship's Flour Room and vegetables into the Veg Room. On top of all that there would be over one thousand cases of beer and soft drinks, plus the bonded stores of cigarettes and duty free spirits. And while all this was being carried onboard, three meals a day had to be produced and served up. Eventually the Catering Staff managed to get ashore, only to find that all the bars shut at 6 o'clock at night, as it was a bit primitive in Australia in the late 50s and early 60s.

I really don't remember much about Australia on my first visit. What I remember about Fremantle was it looking like a cowboy town, with its wooden-style buildings, so I was a bit disappointed. But in later visits I noticed how the country changed to be very modern, particularly in the cities of Perth, Melbourne, Sydney and Brisbane, to name a few.

Sometimes you couldn't get ashore, if it was a fast turn round, as ships cost money when in port. After discharging the cargo of crude oil in Australia, it would be back up the Persian Gulf to load the next cargo. While you were discharging cargo, you would also be taking on Bunkers, for the running of the ship's engines, as you could be at sea for weeks at a time.

The next cargo that was loaded was to be taken round to Rotterdam, back through the Suez Canal, via the Mediterranean Sea, to Europort in the Netherlands. After Rotterdam, the *British Bulldog* loaded the next cargo to go across the Atlantic to New York.

The crossing of the Atlantic would normally take about 10 days at a speed of 14 knots, but on this crossing it took the ship 14 days as we must have hit the back end of a hurricane. It was the worst weather I had been in; sometimes all you could see was water in a deep trough, then you would be riding the crest of a wave. By this time I must have got my sea legs and adjusted to sea life, as I wasn't sick at all despite this ship of 28000 tons loaded being tossed about like a matchstick. It was certainly frightening at times. Prior to arriving at New York, you would have to notify the Port Authorities of all personnel onboard the ship, and their rank, name, date of birth, etc, before being allowed to disembark for shore leave in the USA. You were then given a shore pass which had to be handed back prior to the ship departing US territorial waters.

On arrival, the Port Authorities would board with US Customs and various other dignitaries. At the time everyone on board ship had to see the US Immigration, Port Health, to be examined and check that no-one was infected with any form of venereal diseases. This meant you had to flash your private parts, which was a bit degrading and aptly named "Short Arm Inspection". While we were there, about 10 of the ship's staff, including myself, hired a limousine. It cost about 10 US Dollars each – about a week's wages for me at the time – but it was the best way to see the sights of New York. We were all in awe of the height of the skyscrapers, Staten Island, Broadway, Jack Dempsey's Bar, just to name a few of the sights.

After leaving New York, the ship sailed down to Venezuela to load a cargo of crude oil to take back to the UK.

We sailed up the Maracaibo Lake and I have never seen so many oil wells pumping oil in a river.

After leaving Venezuela, we sailed back across the Atlantic; thank God, a calm crossing that time. The ship had orders to proceed to Milford Haven in Wales to discharge the cargo.

Everyone was in high spirits as we had been away from home for about seven months, and the majority would be ready for shore leave.

On arrival at Milford Haven, a coach would take us to the nearest main line railway station, which was Swansea, and then we could make our way home by train.

We left the ship at 12 o'clock lunchtime and I didn't get home till next day at 07:30hrs; after getting off the bus, I passed my sister heading off to work.

CHAPTER 4

After being on leave for five weeks since signing off the *British Bulldog*, I had asked the Shipping Federation if there would be a possibility of joining a cargo boat. I was sent down to Victoria Docks in London to sign on a ship – owned by a company called Alexander Shipping Ltd – named Motor Vessel *Tewkesbury*, which was going round to Liverpool on a run job, on the 15th May 1962.

The ship's crew signed off, and then re-signed on for a voyage down the South American Coast. After the ship was loaded in Liverpool, we had a deck cargo of two Aberdeen Angus bulls, which were being imported into Argentina for breeding purposes, and two horses. These bulls had a value of £25,000 each, and a specialist cattleman accompanied the voyage down to Argentina to look after them on the three week journey.

Prior to leaving Liverpool, sheds were assembled on the main deck for the livestock, to accommodate their passage to Argentina. The cattleman, who was designated to look after the Aberdeen Angus bulls and horses, would be onboard until the stock was unloaded at Rosario, up the River Plate.

His priority was the wellbeing of the stock until the unloading at the discharge port. Although feeding and looking after the live cargo was his specific duty on the passage down there, at times he would partake in menial tasks on deck with the crew members mainly out of boredom and to pass the time.

Alexander Shipping Co. Ltd was a very progressive company when it came to the accommodation; all the ship's cabins were single berth with a sink in each, which was unusual at the time. The first port on this voyage was Montevideo in Uruguay. The good thing about being on a cargo boat was that the ship would be in port for about a week, because unloading of the cargo was at a slower pace.

After leaving Uruguay, the ship sailed to Argentina to the port of Buenos Aires. On arriving in Argentina, it seemed ages before all the Port Authorities were dealt with as everyone needed an Individual Certificate to get ashore and there was a vast amount of details on it, as illustrated on the form.

The ship could be in Argentina for about four or five weeks, discharging various types of cargo and then back loading. The ship also sailed up the River Plate to two more ports, Rosario being one of them, to unload the Aberdeen Angus bulls and the horses.

After Argentina, the ship returned to Montevideo and loaded more cargo for the voyage back to Liverpool.

The round trip from joining to leaving the ship in Liverpool was just short of three months, but voyages could vary in length.

After leaving M.V. Tewkesbury and signing off the ship on 12th August, 1962, I went home for about four weeks' leave. On this occasion I went to Blackpool for two weeks' holiday with another five mates; that two weeks in Blackpool seemed like an eternity!

The next ship I joined was a coastal vessel, which I had asked for. The ship was somewhat smaller in tonnage, gross tons 963, net tons 375. The M.V. *Darlington*, registered in Hull, was part of British Rail and belonged to Associated Humber Lines Limited, which had about 8 to 12 small coastal vessels.

I joined the ship at Goole – about two-and-a-half hours from where I lived. This meant I used to leave home at Sunday lunch, travel to Pontefract, then get the bus from Pontefract to Goole as there was no rail service on a Sunday to Goole.

The ship was what was termed Home Trade, and it left Goole on Sunday evening and arrived in Rotterdam the next day, picked up a pilot, and was tied up by just after lunchtime in Rotterdam opposite the Euro mast.

The ship would be in Rotterdam for about three days, unloading and back loading cargo, for the return passage back to Goole. I would be back home Friday afternoon by rail. Working for Associated Humber Lines had its advantages, as you would qualify for cheap fares on British Rail as one of their employees.

I stayed on the M.V. *Darlington* for just over three months, but there wasn't much chance of promotion on these types of vessels. The total personnel on board the ship was only eight. My Superior on board was Cook/Steward, so there wasn't much chance of promotion.

I left on the 27th December, 1962, and was only home for about six days before being notified to join another BP Tanker down in the Isle of Grain in Kent, this being one of BP's Oil Refineries near Tilbury.

The ship I was to join was SS *British Victory*, signing on 3/1/63. Getting to the ship at the time was a huge problem, as it was one of the worst winters down on the South East Coast.

After leaving port, we sailed down to Nigeria to load a cargo of crude oil. On the way there, as we were crossing the International Date Line, it was suggested there would be a Crossing the Line ceremony.

This ceremony apparently dates back many years, so a few days prior to the event, our ship was a hive of activity, getting the costumes, etc, ready.

See photographs, with King Neptune and wife and the rest of the rag-tag crew.

Crossing the Line ceremony

All personnel who had never before crossed the International Date Line would be rounded up and, when caught, a special ritual would be carried out by King Neptune.

As the *British Victory* had a small swimming pool, a plank was attached, and each person caught would be tarred and feathered, the barber would go through the motions of cutting their hair, and then they would have to walk the plank into the pool.

After the ceremony, King Neptune would read the ritual to the crew member, to say that on that day the person had officially crossed the International Date Line and a scroll would be given to each one as evidence of its validity.

All ships didn't perform this ceremony and, lucky for me, I had crossed the Date Line on a previous ship.

When we'd loaded down in Nigeria, which was normally a submarine pipeline and tied up to buoys, the ship sailed back to the Isle of Grain in Kent. After being on the *British Victory* for about three months, the ship's crew was signed off but you would be asked if you wanted to re-sign. At the time I had just over two years' sea service and I was asked if I would like to re-sign. I asked the Chief Steward if it was possible to be promoted to Assistant Steward, and he agreed. At the time there was an abundance of sea-going staff in this rank, with quite a few passenger liners being laid up or scrapped, so I was fortunate to be promoted.

After leaving port, we headed round to the Persian Gulf to load a cargo of crude oil, via the Suez Canal. Being promoted to Asst/Steward meant I was doing the duties of Mess Room Steward, which basically meant running the cafeteria style of service to crew members from *bain-marie* containers, washing crockery and cutlery, and maintaining the cleanliness of mess rooms and certain public toilets onboard.

There were some characters on board ship. I remember a bloke called Tim Lynch from Dundee, who was a fireman in the Engine Room. He told me about an incident one day on another ship where he'd shared a cabin; he had had the bottom bunk and his mate the top bunk. Tim was only about five feet tall. He and his mate had gone ashore and got split up, so when he came back to the ship his drunk mate had climbed into Tim's bunk. So Tim got into the top bunk. When he was called for his watch duties, Tim thought he was in the bottom bunk and threw his legs over and dropped about five feet! He said he had been drunk when he woke up, but sobered up when he hit the deck.

In certain months, when you were trading up the Persian Gulf as it was called then, the heat at times could be unbearable. There was no air conditioning and very little breeze, so some nights you would put your mattress out on deck to get a decent night's sleep.

Sometimes the metal bulkheads were so hot you couldn't put your hands on them. Inside the accommodation, you would start work at six and by seven you would be soaked with sweat; you had to take salt tablets to put the salt back into your body or you could dehydrate. It was a Godsend when you left the Gulf area.

After about three-and-a-half months, the ship berthed on a couple of occasions and then sailed back to the Isle of Grain, where the crew signed off. But I re-signed again with the rest of the new crew. I felt a bit of loyalty as I had been promoted on the previous voyage and believed it was payback time. There was an inclination that the ship would shortly be going into dry dock in the latter part of the year, so I reckoned another couple of months at sea was no hardship.

Eventually the ship did go into dry dock up in the North East, at South Shields. This had been my longest time away from home – just over nine months, with signing on and off on three occasions.

While the ship was in dry dock, a skeleton crew was kept onboard to run the ship. I was asked if I would like to sign a contract with the BP Tanker Company, which I agreed to do, as promotion could be quite rapid because many shipping companies were after their own staff.

After leaving the *British Victory*, I was due about six weeks' leave, due to the length of time I had spent on the ship.

CHAPTER 5

The next BP tanker I joined was named the *British Guardsman*. At the time, in the early 60s, BP was a large tanker company, with over 100 ships and another 12 being built or on order. The tonnage of these ships varied in size from just over 8,000 tons to 70,000 tons, dead weight tons. There was also a subsidiary company, named BP Clyde Tanker Company Limited. They varied in the type of cargoes they carried, the majority being crude oil tankers and the remainder being product oil tankers. They had crude oil depots in Finnart near Glasgow, and a refinery in Grangemouth near Edinburgh. Crude oil would be discharged in Finnart and then pumped over to Grangemouth to be refined.

There was another crude oil depot in Milford Haven, and a refinery in Swansea in Wales. And the main one was down in Kent, at the Isle of Grain.

BP named their ships in classes, as in regiments – Br *Hussar*, Br *Guardsman*, Br *Cavalier*, etc. Or tree boats – Br *Fern*, Br *Willow*. Or bird boats – Br *Gull*, Br *Kiwi*, Br *Curlew*. Apparently, one of the boats being named at the time of launching should have been named Br *Thrush*, but the name was pronounced Br *Trust*. Once a ship has been named, that's the name it keeps. Whether this was true or not, I don't know.

The Br *Guardsman*, which I had joined on the 29/11/63 at the Isle of Grain, sailed down to Nigeria and, being one of the company's newer vessels with all aft end accommodation, it was the first ship I had sailed on with air conditioning accommodation. The ship had left port empty, apart from ballast, because after discharging cargo the tanks onboard were being washed out. Going through the Bay of Biscay, the weather was atrocious. It was Christmas Eve, and we were getting ready for Christmas lunch when the ship rode a wave and came crashing down. I thought we had hit something as the ship shuddered, with cups and crockery flying all over the place. In the galley, pots were thrown about in disarray. It was quite an experience before things calmed down.

Steam Vessel "BRITISH GUARDSMAN"
Length 760 feet Breadth 97 feet Deadweight Capacity 52,150 tons
Owned by BP TANKER COMPANY LIMITED
(For list of vessels at August, 1963, see over)

My job on the *British Guardsman* was Engineers' Steward, which involved cleaning the engineers' accommodation, cabins and serving the Officers in the dining saloon.

It was the same job as Mess Room Steward on my previous ship, Br *Victory*, but different duties. After being promoted to Assistant Steward, the wages also increased from £14 a month to £36, with annual increases.

At the time, these ships seemed enormous in size. The Br *Guardsman* was 760 feet long, and 97 feet breadth. But, compared to ships now, it was only small.

The Br *Guardsman* had been built on one of the shipyards on the River Clyde, in Glasgow, and had to go back for the ship's guarantee dry dock. First, the ship has to have its tanks cleaned out and the slop tanks dug out, and this was done in the port of Greenock, prior to going into dry dock for repairs. After being on the Br *Guardsman* for six-and-a-half months, I was being relieved of my duties and heading home on leave.

My next voyage after my leave was on the SS *British Signal*, which I joined at the Isle of Grain in Kent. There wasn't much entertainment at the Isle of Grain; it was quite an isolated community, apart from a small village with a couple of shops and two pubs. I remember talking to the landlady one night and she told me that when she had been in the Armed Forces, she had got herself in the family way, and been demobbed. Entered on her Discharge Book were the words, 'self-inflicted wound'.

Joining the *British Signal*, I was promoted to Second Steward, a rank above Assistant Steward. The main duties of 2nd/Steward were looking after the Captain's and Chief Engineer's accommodation, calling in the mornings, tea/coffee breaks, sandwiches at night, service at meal times from *bain-maries* to Saloon Stewards, service in the Officers' Dining Room. As time progressed, ship life became more bearable, although some shipping companies were better than others. Shell and BP, in the late 50s and early 60s, had a bad name in relation to accommodation and catering onboard. ESSO Oil Company had a good name; their abbreviation was eat, sleep, shit and overtime. BPTC was better, times, coming.

BP progressed as new tonnage was introduced, and their catering standards vastly improved as more and more staff signed contracts relating to job protection and security.

While on the Br *Signal*, the ship went into dry dock in Falmouth, which was unusual. We had to sign off the ship and practically everyone re-signed back on. We weren't long at dry dock, so the problem in the Engine Room wasn't too complicated. The ship was at the time trading from the Persian Gulf to North West Europe, and I signed off at Antwerp in Belgium.

While home on leave, I was approached through the company and asked if I would like to attend South Shields Marine and Technical College on the Tyne. At that time BP had two colleges, where they would put their staff through a 12-week course in practical cookery/ bakery and a theory test. Apart from the one in South Shields, there was also a course in South Wales, at Landarcy Refinery near Swansea.

The course in Wales was sometimes called The Marriage Bureau, as quite a few of the trainees had married female staff who worked in the canteen at the refinery.

The course at South Shields was to start in April 1965. There were 12 students, from various parts of the UK, selected to attend.

On arriving at the college, students had to enrol. I had met one or two of my fellow seafarers who were on the same course, and one of them had asked where I was staying as he lived quite locally, at a place called Howden on Tyne. I told him that I hadn't found any lodgings, so out of the blue he phoned his mother up and asked her if she wanted a lodger. I finished up staying with him and his family for 12 weeks, which was a Godsend, as I hadn't a clue where to start looking for lodgings.

The family I stayed with were named Wilkinson. Wilf was their son; I had sailed with him on one of my previous BP tankers.

Mr and Mrs Wilkinson were down-to-earth; you could have said it was just like being at home. They had an older son who was married, and a daughter who was at school at the time. Mr Wilkinson worked in one of the local shipyards, which I think was Swan Hunter's, as did a lot of the locals. It was one of the major types of employment at the time. After meeting the family, Mrs Wilkinson said her son paid her £3.50 a week and that was what she would charge me for my week's lodgings if that was okay by me. I was highly delighted and thanked her for allowing me to stay with their family.

The catering course at South Shields College should have been for 12 weeks, but for some reason it had been reduced to 10 weeks and we had to make the time up by doing two nights a week of extra lessons.

The course lecturer was a man who had served his time in the Royal Navy, working his way up through different institutions to course lecturer.

He was very strict, but also fair, and wouldn't belittle you if you made a mistake. The course was basically based round the cookery and theory of Ceserani and Kinton, two well known chefs in the catering industry.

The course was set up to go through different stages of a menu – soup, fish, meat, desserts, etc. One week would be *hors d'oeuvres* in different types and shapes, next would be basic sauces, and so on, with practical and theory in butchery and filleting of different types of fish.

Prior to starting the course, we had to kit ourselves out with catering clothes, cook checks, aprons, jackets and hats, and a set of knives for their appropriate uses.

Once the course started, we had to make our way to the college from Howden on Tyne to South Shields, via the Jarrow Tunnel.

Luckily there was another lad who used to pick us up at the tunnel and give us a lift to the college.

We soon got into a routine, and things started falling into place with what was expected.

While in the North East, one of Wilf's mates, Ronnie, was friendly with a group of musicians, and we sometimes went to see them perform at the weekends. They were called the "Bodysnatchers". They never made the big time but we got to see them a few times in the surrounding area. One of the big weekends on the Tyne was when the Tyne Moor Feast was on. It was one of the biggest fairs I had ever been to, the only problem was we had been drinking Newcastle Brown Ale so I don't remember much about that night. It was deadly the effect it had on you, so after that night, I stayed away from brown ale.

Mrs Wilkinson used to do her baking on Sundays, so Wilf and myself on a few occasions would give her a break and do the baking, using our practical skills to gain some experience for when we went to sea.

It was a memorable time in the North East; they were very friendly people. I remember Mr Wilkinson, Wilf's dad, asking his wife if she'd got his bait ready for the next day. I thought he was going fishing, but it was his lunch box for when he was at work.

All good things come to an end, as did the course at the Technical College. When the time came round to sit the examination, all the candidates passed with flying colours.

We had a night's celebration with all the lads on the course, and wished everyone well on their next voyage as 2nd/Cook & Bakers.

The saddest part for me was saying goodbye to the family I had stayed with for nearly three months. They had made me very welcome at their home and I will always treasure their hospitality.

Wilf and Ronnie

CHAPTER 6

After a couple of weeks at home, I was sent to Southampton to join the *British Trust* as 2nd/Cook and Baker, trading around the UK and Northern Europe.

The biggest adjustment I found was that at South Shields College I had been cooking for four people, yet onboard ship I was cooking for approximately 60. It took a while to get used to portion control.

But I must admit the Chief Cook was very helpful at the time; he was a Scot from Bo'ness, near Grangemouth.

My first job was getting the dough made for the bread. I used to make a 40lb batch, which would make 10 x 4lb loaves, and 60 bread buns. This would take about four hours from making the dough, proving it twice, then baking it in the oven.

This went on six days a week, and on Sundays I just made bread rolls. If the vessel was in UK ports, the Chief Steward would order bread locally, so that gave me a bit of a break until back at sea.

I had joined the Br *Trust* on the 3rd August, 1965, then the ship had orders to go to Newcastle for dry dock, which I thought was a bit uncanny; I had just spent three months on the Tyne, and two months later I was back there.

I did visit the Wilkinsons while I was up there, but unfortunately Wilf was away at sea.

After leaving dry dock just outside the Pier Head on the River Tyne, the ship broke down but the engineers onboard fixed the problem. The problem was the ship's engine – a Burmienster and Wain, 6 cylinder engine. The first time I had seen this type of engine was scary, watching it pumping up and down, driving the main shaft/propeller.

After leaving the Tyne, the ship went to load a cargo at the Isle of Grain and discharge it in Grangemouth, Scotland.

After leaving Scotland, the Br *Trust* had orders to proceed round to the Persian Gulf to load a cargo.

The only problem with the ship was it just kept breaking down. After leaving Grangemouth, the ship struggled round to the Mediterranean Sea, approaching Malta. The company decided to put into port at Malta and we were there a week till they sorted out the problem.

Leaving Malta, we sailed to Port Said for transit through the Suez Canal, but more problems arose and we were at anchor for a few days with more repairs. Sometimes, prior to passage through the canal, the ship's management team would hire a squad of locals and have the whole main deck painted to improve the appearance of the ship; it wasn't as costly as getting the ship's staff to do it.

After passage through the Suez Canal, you enter into the Red Sea.

There were still problems with the engine cylinders, which had to be changed, and on approaching Aden, the ship pulled in for more repairs. While I had been on leave between voyages in the early 60s, quite a few young men had enlisted in the Armed Forces. I had mates who had joined up, and we had enjoyed a few nights out together when they were home on leave. One lad, who had been posted out to Aden for a nine months' tour of duty, was called Louis Stocks. During the 60s there had been an uprising in Aden, so the British Forces had a base there. While the ship was in Aden for repairs, I had met one or two Army blokes and asked them to enquire if Louis was at the base in Aden.

Apparently what used to happen on the nine months' tour of duty at Aden, the first three months would be to acclimatize to the conditions, then the soldiers would be sent up country to root out the insurgents. It seems a similar type of warfare to that which the British Forces are fighting out in Afghanistan at the present time.

Louis had been posted up country in the Yemen mountainous region, so I didn't get a chance to see him then, but we had a few pints later on back at home.

After leaving Aden, the ship progressed round to the Persian Gulf. We eventually arrived at our destination at Bandar Mahshahr in Iran, to load a cargo for Durban in South Africa.

When the cargo was loaded, we set sail down the river from Bandar Mahshahr. Due to the tide being so far down, we had to anchor to get over the sand bar. I had been making about 60 cream horns for the afternoon tea for Officers and Crew Members. I went out on deck for a smoke, to let the catering boy scrub the galley deck down, and let the cakes cool prior to putting the jam and cream in.

While I was having a smoke, the catering boy came out on deck and said he had knocked the cream horns off the bench. I thought he was joking, as he was a Sunderland lad and a bit of a prankster – until I looked. There were cream horns floating in soapy water all over the galley deck. If I could have caught him, I would have strangled him. It had taken me an hour-and-a-half to make them, so I finished up making 60 scones in my own time.

After leaving the anchorage, we set sail for Durban, to discharge the cargo. On the way there we had a few more breakdowns but finally reached the port of Durban, just short of two weeks after discharging the cargo. Local representatives were then employed to sort the problems with the ship's engine.

We had some time off to see Durban. One day, going through the docks area, there was a squad of prisoners from the local prison, working on the railways; all of them were chained up in a gang. I couldn't believe it. It was like a scene out of a film.

It was my first experience of seeing segregation in an apartheid regime. Down on the beaches there were whites-only beaches, ditto for the blacks. The buses were the same, and there were no blacks in the whites-only bars, except those working there for a living.

After storing the ship in Durban prior to departure, we set sail back up the Persian Gulf where we had orders to load a cargo at Bandar Mahshahr and proceed back to the UK.

Sometimes it made you wonder how the oil tanker companies made money. I was told later on in my seafaring years that it was the only part of oil companies that made a loss. All the money was made in the exploration of oil, and in the refineries and by-products.

We were still having breakdowns, though not as frequently, and eventually we arrived at the Isle of Grain in Kent, where most of the ship's staff were relieved of their duties and allowed home on leave. I had done nine months on the *British Trust*, so I was looking forward to some time at home.

After being home for just over two months, I was to join the M.V. *British Gannet*, another one of the company's bird boats, in the Isle of Grain in Kent. This was one of the main ports for this class of ship; nearly all the bird boats, as they were called, were around 15,000 dead weight tons. They were all product carriers and ideal for getting into smaller ports in the UK, Northern Europe, and worldwide.

After joining the Br *Gannet*, we did coastal trading for about three weeks, then the orders came through to proceed to the Persian Gulf to load a cargo again at Bandar Mahshahr.

Although we had orders to load in Iran, sometimes you wouldn't be told where the cargo was to be discharged until a later date. The passage from the UK to Iran would normally take about 21 days, via the Suez Canal. Fortunately the Br *Gannet* didn't have the same problems as the Br *Trust*, and wasn't breaking down as many times. After loading at Bandar Mahshahr, the ship was to discharge in various ports on the Indian Sub-Continent. The first port of call in India was to be Bombay, now called Mumbai.

Prior to arriving at Bombay, the Chief Steward would be required to complete a Customs Declaration Form to be filled in by every one of the ship's Officers and crew members, as in every foreign port of call visited. Four copies had to be completed, listing cigarettes, spirits, radios, cassettes and watches, plus any foreign currencies.

Apart from the cigarettes and spirits, everything else must be on board the ship when leaving Indian territorial waters, because the ship was visiting five different ports in India.

The top copy of the Customs Declaration Form was sealed, and this copy travelled around the Indian Coast with the ship. It was known as the Circulating Copy.

Bombay was an eye-opener for me, as it was my first time there. Poverty seemed to be everywhere; if you walked down the streets, beggars were all around, asking for money. There was destitution wherever you went.

After leaving Bombay, the Chief Cook had to go into hospital and was scheduled to rejoin the ship at a later date, on the Indian Coast. The next port of call was Cochin, and the Chief Steward would be in the galley for certain times of day; he had a Chief Cook's Certificate as I wasn't fully qualified. Before we arrived at Cochin, another Customs Declaration Form had to be filled in for the Port Authorities there.

On the way down from Bombay to Cochin, which took about four days, one of the Hobart parts snapped. This was one of the parts which was attached to the machine whilst making the bread dough.

There weren't any spare parts to fall back on, so the only alternative was for me to make 40lbs of bread by hand. It wasn't an ideal situation as the weather at the time was quite warm, and it proved quite strenuous until I got used to it. A new part was ordered, but you never knew how long it would take to get it out to the ship.

Going ashore in India, especially if you were sampling the catering establishments, could be quite daunting, and quite a few crew members went down with diarrhoea and sickness, including me.

After leaving Cochin, the ship was to proceed to Madras, which was round the coast. The same procedure had to be adhered to on arrival at Madras, with regards to the Customs Declaration Forms, and the same again when the ship arrived at Calcutta. So far none of the Customs Officers who had visited the ship on the Indian Coast had opened the sealed circulation copy, for some unknown reason.

We arrived at Calcutta, and I have never seen so much activity tying and securing a ship to a berth. I didn't find out till later on that the river into Calcutta is very tidal – similar to the River Severn with a tidal bore – which was why huge chains, wire ropes and other ropes were used to secure the ship to the jetty.

While we were at Calcutta, the Chief Cook, who was from the Orkney Isles, had rejoined the ship, all back in good health. It was his first trip as Chief Cook. The galley on this ship was fitted in the same way as the rest of the bird boats, with oil-fired stoves and ovens, which were great when they were working, but were sometimes terrible to flash up.

One of the firemen used to get them going in the morning, but occasionally there would be a flash back with a carbon build-up during the day, or they would be out and you would have to relight the burners. Sometimes the cook would have his face in front of the opening, relighting the burner, and I would tell him to remove his face in case there was a flash back with a flame. I used to dread the consequences of what might happen.

I didn't go ashore while the ship was at Calcutta, but I remember one of our local doctors at home, who was originally from there, saying to me that it was one of the dirtiest cities in the world.

On leaving Calcutta we were to go back round to Bombay empty, and then reload the ship to take a cargo of oil up the Gulf of Kutch to a port named Kandla, in Northern India.

It was about five to six days from Calcutta round to Bombay; you don't realise the size of India until you travel round the perimeter of the continent. The ship would then be in port at Bombay for about 48 hours, prior to loading for Kandla.

Manoeuvring the ship in the Gulf of Cutch was a real feat of seamanship, as this was also a fast flowing river. The ship was taken up by the pilot at a certain time of the river's flow, secured on buoys, and then a submarine pipeline was picked up and the cargo would be pumped ashore.

This was also when the Indian Customs officers swooped. Prior to arrival at Kandla, a Customs Declaration Form had been completed as usual. But this was when the original circulating copy was opened and everything declared on it, apart from cigarettes and spirits, had to be accounted for.

Everyone's personal customs declared must be on the original copy and also on the last declaration submitted at the Port of Kandla, the last port of call.

Everyone's cabin was checked for any discrepancies, and if any extras were found that differed from the Customs Form, they would be fined for the selling of watches, radios or cassettes, in or on the Indian Coast. Quite a few of the crew were fined. When checking my cabin, I was asked what I had declared, which I told the Customs officer. He then asked me if I had declared any foreign currency, which I had – £10 sterling. This had been on the original copy but, for some unknown reason, not on the last one, so the money was taken. I had intended to keep this money for when I eventually was paid off the ship. The 2nd Officer, who came round with the Customs Officer, told me not to worry, and a couple of hours later I was handed my £10 note back, thanks to him.

I think everyone was glad to get off the Indian Coast and back to sea, and put it down to experience. After departure from Kandla, the ship was to proceed back up the Persian Gulf to load a cargo at Bandar Mahshahr in Iran.

After loading the ship, we had orders to proceed to North West Europe with no definite port of discharge. So it was back through the Suez Canal. The canal itself was continually being dredged to widen and deepen it, as ships were getting longer and much larger.

We went out through the canal into the Mediterranean Sea and then received the orders to proceed to Swansea to discharge the cargo. This would mean leaving the ship to go home on leave. As we were on contract to the BP Tanker Company, the voyages were becoming slightly shorter trips – usually four-and-a-half to five months at sea – but it also meant our monthly salary would be paid into our designated bank.

Normally when arriving back in the UK, the ship would be a hive of activity with Customs, Port Health, Agents, etc, not only discharging cargo but also loading ship's stores. There would also be shore staff to load the stores, as ship's staff would be leaving and new staff joining. Laundry had to be put ashore and returned prior to the ship's departure.

I left the Br *Gannet* at Swansea in Wales to go home on leave on 29th November, 1966. I was to get extra leave on this occasion, as on the previous voyage I had been called back early due to being short staffed.

That meant that on this occasion I would be having a Christmas at home with the family.

Sometimes it could be awkward trying to arrange holidays when home on leave. I did manage to get down to Wembley to the Empire Stadium, as it was called in 1966, to the Rugby League Final. It cost 20 shillings to get into the seat that was allocated, or £1 in today's money. I can't remember who was playing, but it was a good five day break with the West End Club, at Hemsworth.

CHAPTER 7

The next BP Tanker I was to join was the *British Ensign* at Rotterdam. During the mid-60s and early 70s, you could practically be sent anywhere in the world to join a ship. So you had to have an up-to-date passport and vaccination certificates. The *British Ensign* was quite large by the standards at the time, 70,000 D.W. tons; she and the *British Mariner* were both the same tonnage and the largest oil tankers in the BP fleet.

In the late 60s things were changing at sea, and Officers' and crew bars were being introduced on board ships. Cash was being used more for the likes of paying for bonded stores and slop chest items, toothpaste, sweets, cigarettes, tobacco, etc. Everything was paid for by cash with a weekly sub issue, and cash was used in both the bars.

Films at sea were becoming more frequent, and videos, and these could be exchanged at certain ports of call throughout the world. With the introduction of larger tankers, some were restricted from going through the Suez Canal, including the Br *Ensign*. Passage would instead be round the Cape of Africa and up to the Persian Gulf to load a cargo; this would take about two months for the round voyage.

Passage round the Cape would be interrupted by one or two helicopters coming out to the ship from Cape Town, bringing fresh vegetables, fruit and salads, mail from home and the exchanging of films.

The Chief Steward on board ship was from Doncaster, just 14 miles away from where I lived. While on another B.P. ship, he had suffered a burst ulcer, just off the US Coast, and a helicopter had been sent by the US Coastguard to take him to hospital. He told me that after he came round and recovered, the doctor had said he had come into hospital a full-blooded limey, but was going out a full-blooded Yank, thanks to a blood transfusion of almost eight pints of blood they'd pumped into him.

Joe Fouweather was his name, and he was one of the Chief Stewards who encouraged me to go for my Chief Cook's Certificate next time I was on leave.

On the way round the Cape there was an abundance of albatross, which you would normally see south of the Azores, or there would be dolphins or porpoises up alongside the ship. Occasionally there would be whales, or flying fish would land on the deck.

After loading the cargo it would be back round the Cape, to either Rotterdam, Finnart in Scotland, or Milford Haven in Wales.

We had a couple of cargoes in between, down in Bonny, in Nigeria.

During the trip on the *British Ensign*, I had made up my mind to apply for my Ship's Cook's Certificate next time on leave, as I had then chalked up two years as 2nd/Cook & Baker.

I left the *British Ensign* at Milford Haven on 16th August, 1967. Not long after arriving home, I notified head office I wanted to apply to Hull Nautical Cookery School to take my Cook's Certificate, and they agreed. The course would be for six weeks' duration. I wasn't as lucky at finding accommodation this time. I stayed in the Seamen's Mission in Hull and, although it wasn't The Ritz, it was adequate for the six weeks I was there.

I think at the time there were only three of us doing the Cook's Certificate. The course itself didn't seem to be too well organised; you would be given certain tasks to do and told to just get on with it.

The food that was produced up till lunchtime was then eaten by the Shipping Federation Staff or Union Staff, with a small charge going towards the cost of preparing it. After lunch there would be a clean-up of pots and pans. With the practical work done, it would then be a couple of sessions of theory. I remember one day, after we had prepared lunch and we were having our break, there was a disturbance outside. We went to have a look and there were two women knocking seven bells out of each other, outside a notorious ale house in Hull. Somebody must have phoned the police and in a matter of less than 20 minutes, four police cars appeared, a police van, and a riot van. Eventually two of the policemen went over and said, 'We don't mind you fighting, but we expect you to break it up when we come.' They then threw the women in the van and carted them off.

That was the highlight of the six week course at Hull Nautical Cookery School. Prior to finishing the course, there was an exam to take. I passed, and was issued with a certificate to say I was eligible to perform the duties of Chief & Ship's Cook in the Merchant Navy. I went home to finish my leave. So, from signing a contract with BP Tanker Co. Ltd in 1963, I had risen in rank to Chief & Ship's Cook in just four years.

The next BP Tanker I was to join was the *British Willow*, one of the Tree Boat Classes, on the 26th October, 1967.

The benefit of reaching the rank of Chief Cook was you received the same leave as the Officers on board ship. So for doing a four-and-a-half months' tour of duty, you would be due about two months' leave, plus there was an increase in wages to go with the responsibilities of the job. I joined the *British Willow* at the Isle of Grain, and we were to trade around the UK Coast, North West Europe and up the Baltic for a couple of months. On the 2nd January, 1968 the Articles of Agreement were changed, as too many staff had been engaged. New Articles of Agreement were signed and, apart from staff going on leave, the remainder re-signed on. We then were told to head for the Persian Gulf, receiving further orders when nearing the destination.

As we approached the Persian Gulf, we were told to load at Bandar Mahshahr, and to discharge the cargo at Kwinana, Western Australia.

Before leaving the United Kingdom, two firemen had signed on the ship and it was a bit of a coincidence, but they were two brothers who were from the next village to where I was born, Fitzwilliam. I didn't know them at the time, but I got to know about them as they became shipmates for the next four months. Ronnie and Joe Hobson were their names. They had another brother, called Denis, who was also with BP Tankers.

When we eventually arrived in Western Australia, things had changed a lot since my first visit to Kwinana. The first time, the pubs used to close at six o'clock at night, so if you wanted a drink while ashore, you had to have one during the day time. But the licensing laws had changed the second time I visited, to 10 o'clock at night.

While we were in Kwinana, the two brothers I have mentioned had asked the Chief Engineer if they could have the weekend off to go see a friend they had worked with down one of the mines in the UK. He granted the time off, and they hitchhiked 100 miles inland to see their friend.

The reason the ship was in Western Australia for so long was that only so much of the cargo was being discharged at Kwinana, and the ship was what is known as "back loading" a different fuel oil. This was one of the ways a ship could actually get to trade on the Australian Coast. In the Australian Merchant Navy, they had a very strong union and normally only Australian ships traded on their coast.

After leaving Kwinana we went round to Adelaide, to discharge a parcel of the cargo, then on to Melbourne. While the ship was in Adelaide, the Chief Steward had ordered some sheep brains, which came in frozen form. One day they were on the menu. It was something I had never seen before, never mind cooked. After defrosting them, I was washing them, and – excuse the pun – the 2nd/Cook came over and said, 'Is that what you call being brain-washed?'

I have never cooked brains in so many different ways – braised, sautéed, fricasseed and poached.

After leaving Melbourne, it was round to Sydney to discharge another part of the cargo, then up to Brisbane to offload the remainder. When we left the Australian Coast, we were supposed to go back up the Persian Gulf; at the time the Suez Canal was closed due to the Arab-Israeli Suez War. I thought it would take another two months at least, as it meant the ship would have to go round the Cape, but for some unknown reason the ship was diverted to Singapore to go into dry dock.

This meant the majority of Officers and crew would be relieved of their duties, paid off and sent home on leave.

At the time I had served seven months on the *British Willow*. When we were told we were going home, I was doing cartwheels down the alleyway. It meant I was due about three months' leave on this occasion, so I went to Lido De Jesolo on holiday with my elder brother and two of his workmates. It was on the Italian Adriatic, just below Venice, and we had two weeks' holiday there. It wasn't to their liking, as their previous holidays had been in Spain which was a bit more vibrant; Italy was more conservative. In fact, we had asked the travel courier if it was possible to get an early flight home, but it wasn't possible so we stuck it out.

CHAPTER 8

The next ship I was to join was the *British Loyalty*, known as the Ity Class. At the time, the BP Tanker Company was going through a transformation and having new tonnage being built throughout the world.

The new tonnage was a huge improvement on their present fleet, with everyone on board ship being provided with a shower and toilet in each cabin, and air-conditioning throughout.

As the majority of staff had signed on contract to the company, working practices changed on board ship. What was introduced was known as GP Manning, which meant if any big jobs on deck needed doing, all the manpower was put to use on deck, or likewise used in the

Engine Room. Weekly meetings were introduced to discuss the work programmes to get the best out of the manpower.

One thing which was done on a weekly basis was the fire and safety drill with mock fires. Stretcher parties were integrated during the drills, lowering the lifeboats to ensure everything was in working order. The GP Manning had the full cooperation of the Seamen's Union and the Officers' Union. With the framework of the new working structure came increases in wages, but less manpower on board ship. The manpower roughly on board ship when I joined the Merchant Navy, depending on the size of the ship, would vary from 50 to 60. With the introduction of GP Manning, it dropped to approximately 35; this would be the same manpower on a 200,000 ton tanker or an 18,000 ton tanker.

Sometimes a full crew would join a ship to relieve the ones on board, and they would be flown home on leave. This could take place anywhere in the world.

The *British Loyalty* I was joining was being built in one of Sweden's shipyards, called Eriksberg, in Gothenburg.

For some unknown reason, I was joining the ship on my own, as so many of the deck and engineering staff were already out there.

The first week there I was staying in a hotel called The Oppallen. The standard of living in Sweden was far superior to the UK and I was glad I wasn't picking up the hotel bill for my stay. It cost about £1.50 for a beer or about £4 for a rum and coke, and that was in 1968.

We used to go down to the shipyard, as there wasn't much we could do apart from get to know the layout of the ship and wait to fit the galley out with the equipment which was in storage.

After over a week, the remainder of the crew arrived at the same hotel where I was staying; it was a welcome sight to see the rest of the catering staff and all the other crew members.

A few days later, we all were sent down to the shipyard to join the ship. It was a bit strange joining a brand new ship and fitting everything out, getting the fridges stocked up ready for sailing and preparing the meals. After departing from Gothenburg, we were trading round North

Western Europe for a few weeks. With the ship being brand new, a Guarantee Engineer sailed with us, checking everything was running to plan. After a couple of weeks he left, and we sailed across the Atlantic with a cargo to unload in Venezuela. A few of the South American countries had volatile types of government; you had to be careful when venturing ashore as, if you took any more than two packets of cigarettes, the armed military would confiscate them.

I have never seen military with so much weaponry; they were armed to the teeth. In Venezuela there seemed to be a lot of corruption, which would have been likely with it being an oil producing country. Apparently some shipping companies, prior to going to Venezuela, would offload a lot of their bonded stores because, upon arrival there, the Customs would confiscate the lot.

From Venezuela we went to Netherland Antilles to load a cargo and then headed back to the UK.

One day on board ship, the Engineer Steward went to see the Chief Steward to say that the vacuum cleaner had packed in. One of the Engineer Steward's duties was to call the engineers prior to going on watch, and give them a cup of tea or coffee. So the Chief Steward thought he would check the machine before he reported the incident to the Electrician. When he stripped the machine down, the vacuum cleaner bag was soaking wet. When he asked how the bag was wet, the Engineer Steward grunted and said, 'Have you ever seen a vacuum cleaner drink a cup of tea?' Apparently, if the engineers didn't drink their tea or coffee, he used to suck it through the cleaner. When the Chief Steward told me at the time, I thought it was hilarious.

We came back across the Atlantic to unload the cargo at Southampton. BP had a small port at Southampton, called Hamble; across the River Solent, ESSO had a huge refinery which was their main depot for the UK.

Whilst discharging the cargo at Hamble 20 weeks stores was also being loaded consisting of beef, lamb, pork, bacon and different types of offal; 12 weeks' fish, likewise types; 9 weeks' eggs; 12 weeks' veg, consisting of fresh, frozen, dehydrated; 8 weeks' fresh fruit; and all the different types of dry provisions. Normally there would be 1,000 cases of beer, soft drinks in cans, bonded stores, cigarettes, tobacco, different types of whisky, rum, gin and vodka for the bars on board.

After leaving Southampton, M.V. *British Loyalty* (M.V. is an abbreviation for motor vessel) was to proceed to the Persian Gulf. By this time I had been just short of four months on the Br *Loyalty*. The Suez Crisis had calmed down and the Suez Canal had reopened after the Six Day War, so passage through was back on track.

As the ship neared the Persian Gulf, quite a few of the staff had been notified that they would be relieved of their duties and sent home on leave, including me.

What normally happened was that new staff would be brought to Ras Al Khaimah (or R.A.K.) in Dubai, then transferred out to the ship at sea, while the original staff would be sent back to Dubai to go on leave.

Senior Staff, the Captain, Chief or 2nd Engineer or Chief Officers, would have a two day handover – theirs was a more responsible job – and be relieved at the loading port.

I was told I would be relieved at Bandar Mahshahr, which was going to be the loading port. I was packed up and ready to go when the Chief Steward came down to my cabin and told me that the Cook, who was supposed to be taking over from me, had missed the flight out to Tehran in Iran, and I had to stay. You can imagine I wasn't too happy at having to unpack my suitcase.

After leaving the Gulf, the ship loaded, we went to Thailand to discharge the cargo. Upon arrival, the ship was crawling with women. Nobody knew where they had come from, but eventually they were all cleared off for safety reasons.

After leaving Thailand, it was back to Aden to load a cargo, which we were to take round The Cape to Angola. The trip from Aden down the East Coast of Africa was stunning at times with the sunsets or the tropical storms. Sometimes there would be shoals upon shoals of fish, with the birds diving down to catch them. It was unbelievable to see the dolphins up ahead of the forecastle, jumping in front of the ship or alongside as we were steaming along.

On this occasion we didn't load any provisions going round the Cape, as the ship would be loading fresh produce at Luanda, the capital of Angola.

Angola was once a Portuguese Colony, but had eventually been handed back. Just before we arrived in Luanda, I was told I would be leaving the ship there and going on leave.

The Chief Steward had ordered the ship's provisions while we were there, and he had ordered 50lbs of crayfish tails. What came was 50 kilos, but they were all alive, so my last job on the British Loyalty was cooking lobsters in the stockpot, prior to freezing them.

After leaving the Br *Loyalty* on the 5th April, 1969, I'd chalked up seven months on the ship thanks to the extra two months since Bandar Mahshahr. At least I would have approximately three months' leave after arriving home.

Sometimes you could imagine why there was a lot of alcoholism and marriage breakdowns in the seafaring fraternity, due to the length of leave, but if I didn't go away on holiday when home I would be out walking or golfing. When I came home, I certainly made up for any leisure I missed when I was at sea.

I don't know to this day why I didn't sit my driving test. I must have thought I wouldn't need a car, being in the Merchant Navy, but it's one of the biggest mistakes of my life. The next ship I was to join was the *British Unity*, which was the same class of ship as the previous ship – another one of the Ity Boats. I think there were about eight of this class built.

The M.V. *British Unity* was being built out in Yugoslavia, as it was known then, at the Port of Split. It was unusual for the BP Tanker Company to have ships built there, but it might have been due to a trade deal between the two countries.

We were there nearly three weeks before we had even seen the ship; somebody must have made a mistake in sending so many of the staff out so early.

The standard of living in Yugoslavia was certainly a big difference to the standard in Sweden, where I had joined my last ship.

Compared to the cost of living in Sweden, Yugoslavia was like a holiday. I don't think at the time tourism had really taken off there.

Before the ship was handed over, sea trials were taken to test all the machinery and the ship's engines. One of the Catering Staff was nominated to go on the sea trials, so the Chief Steward pulled rank, and I had to go.

The ship built in Yugoslavia actually went out to sea for two days, whereas in Sweden the sea trials before handover had been very different.

After the trials, all the crew were sent out to join the ship, ready for sailing. And all the ship's provisions and bonded stores were sent out by refrigerated containers from the United Kingdom. I don't know why everything was sent out by truck, but it must have been less costly. There were about 10 trucks of assorted ship stores, all to be unloaded and packed away in their appropriate fridges and store rooms onboard ship.

We left Split around the end of July 1969 and loaded in Sardinia to take a cargo to Middlesbrough in the North East of England. These new ships were far superior to the older tonnage. The galley was air-conditioned and was a joy to work in when in warmer climates; you could enjoy a good night's sleep without waking up drowned in sweat.

On arriving at Middlesbrough, everyone signed off and the articles were changed from Foreign Going Running Agreement to Foreign Going. After just two days in the North East, we headed back to the Persian Gulf to load the next cargo.

After going through the Suez Canal, we received orders to load a cargo at Aden, and to head for a port in Thailand called Sattahip.

Prior to arrival at Aden, the ship's cargo tanks had to be thoroughly washed and cleaned as the cargo to be loaded was aviation turbo kerosene (A.T.K.).

When we arrived at Aden, US inspectors checked all the tanks, and we found out later that this was because our cargo was for the B52 Bombers, which were bombing Vietnam. Samples would be taken at intervals throughout the loading to ensure that there wasn't any contamination of the cargo.

We left for Thailand and arrived at the US Base to unload the cargo; more samples were taken prior to discharging the fuel oil. US Military officers were onboard at all times during the unloading. I think that people onboard ship were a bit on edge, being so close to the war in Vietnam, but everything went quite smoothly with the unloading. In fact it went so well that the ship was instructed to go back with another cargo for the US Forces, but this time to a different country and base.

The next cargo that was to be loaded was for a port in the Philippines, called Subic Bay.

Aden at that time was a duty free port, so it was a great place to purchase cameras, radios, cassette recorders, watches, etc. There used to be a canteen in the confines of the refinery, where these items could be purchased.

Subic Bay was a huge US Base. Apparently US troops would be sent there prior to going into combat in Vietnam, or troops would be sent from Vietnam prior to going home on leave. Nobody knew what to expect after arriving there, as a sub list hadn't been put round. One or two did go ashore the first day and, once word got round, everyone put in for money to go ashore. It was certainly an eye-opener. Every bar had a group on or some kind of entertainment; it was like an open-air brothel. They certainly looked after their troops before going into battle, or coming back from combat; it was like one big recreation camp.

After leaving the Philippines, we travelled back to the Persian Gulf to load a cargo at Bandar Mahshahr. The cargo that was to be loaded in Iran would be taken to South Africa, to a place called Mossel Bay.

On arriving there, a submarine pipeline was connected to the ship after securing to two buoys.

I have never seen so many hammerhead sharks at one time. All they were doing all day was circling round the ship. I don't think many of the ship's staff ventured ashore while the ship was at Mossel Bay. I know for a fact I didn't; it was a bit eerie to see so many sharks around.

After leaving the South African coast, we were to go back to Bandar Mahshahr to load the next cargo. Prior to arriving at Bandar Mahshahr, I was told I would be relieved there to go home on leave. After the incident on my previous ship, I was a bit sceptical. But fortunately, I was the only one to leave the ship. I had a night in Abadan and then flew to Tehran to catch a connecting flight to Heathrow.

CHAPTER 9

After being at home for just over two-and-a-half months, I was notified that I was to join one of the company's super tankers, which was being built in a Japanese shipyard.

The name of the ship was S.S. *British Explorer* and it was being built in Nagasaki. BP was having eight of these monsters built, but the cost in manpower was huge. One of these ships would do away with four 50,000 ton tankers, plus 3 x 35 seamen. I felt at the time privileged that I had been selected to join one of the first super tankers, but in later years even these were dwarfed. The Br *Explorer* was 215,000 ton, as were the other seven being built; these were known as VLCC or Very Large Crude Carriers. I remember seeing one of these VLCC at the port in Mina Al Ahmadi in Kuwait, and it was just short of a half-a-million ton ship, with 18 men onboard.

The majority of the crew members were joining the ship at the same time, so we left Heathrow Airport at 12am Saturday lunchtime and didn't arrive out in Nagasaki till about 8pm on Tuesday night; it was the longest time I had ever travelled to join a ship.

From Heathrow we flew to Amsterdam, where we had an overnight stop. The majority of the crew joining the ship were from Belfast. Next morning the plane, which was half air freight and half passengers, took off, but had to go back as there were engine problems. A few hours later we left again with one of our crew members serenading a stewardess as the plane was hurtling down the runway. I remember thinking this was going to be some voyage!

After a short stop out in the Gulf, we flew on to Bangkok in Thailand.

Prior to landing, the pilot came on the intercom and said it had been one of the best flights he had flown and thanked all those leaving at Bangkok. The remainder of the passengers staying on were going to Singapore. We had a 12 hour stop in a hotel in Thailand, before joining another plane to Tokyo, via Manila in the Philippines.

After arriving at Tokyo Airport, we were told there was a 12 hour stop – but it was at the airport. Eventually, we got a flight down to Nagasaki, and arrived at the hotel on Tuesday evening. Understandably, it took us a couple of days to recover from jetlag.

We had about a week in Nagasaki before eventually joining the ship; just time to get over the initial shock of seeing it. It was just over 360 yards in length, 56 yards wide, 30 yards depth. The consumption of fuel on a passage from the Middle East via the Cape of Good Hope, on a 30 day voyage, would be 1,371,870 gallons. This is an average consumption rate of 112 gallons per mile or, to put it another way, 18 yards to the gallon.

Getting the ship fitted out and ready for leaving was a major task, similar to the previous two ships I had joined.

There was a big send-off from the Japanese shipyard. The pomp and ceremony prior to leaving the dockyard included a band playing *Auld Lang Syne*, and there were streamers attached from the ship to the jetty as we were pulling away. Once out at sea, though, it was back to the grind.

The British Explorer

The British Explorer

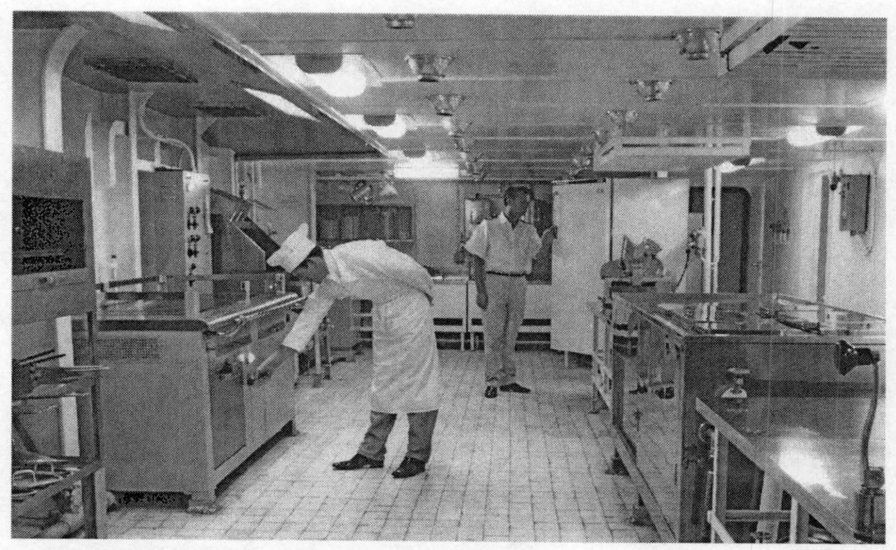

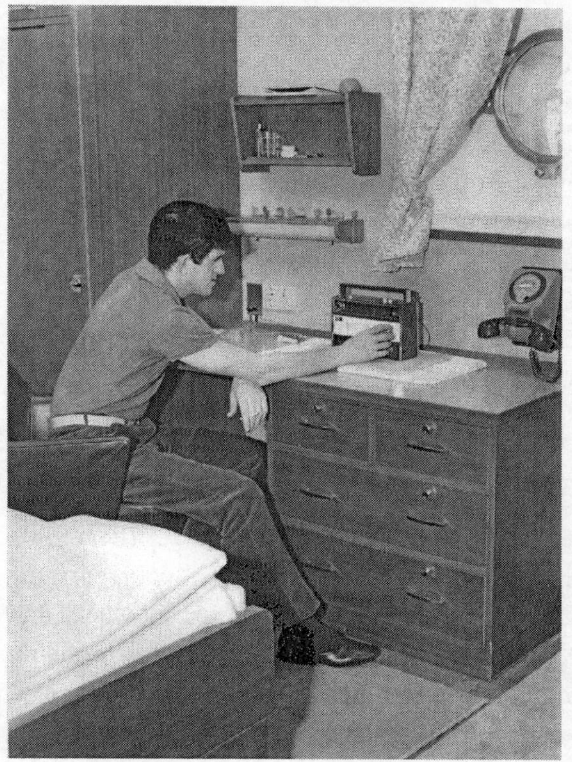

Above: *Myself Chief Cook*

Left: *Myself off duty*

The only problem with these VLCC, as they were called, was where they could load and discharge their cargoes was restricted by their size.

I think the first cargo we loaded was at Ras Tanura in Saudi Arabia. We didn't know till later on that the company had decided to make a film on the way back from the Gulf to North West Europe. So a film crew joined to make a promotional recruitment film, relating to life at sea.

The voyage from Ras Tanura to Europort would take approximately 30 days, round the Cape of Good Hope, as these VLCC were far too big to transit the Suez Canal. I never saw the end product of the film that was made on the passage round the Cape. They did a few shoots in the galley, but were mainly in the engine room, out on deck and doing bridge shots.

We took stores while the ship was steaming past Cape Town; two helicopters came out and unloaded the stores from slings onto the main deck, with mail and a change of films. You wouldn't have thought that ships of this size would roll around, but fortunately they were made to sway and roll with the sea or they could break their back.

After arriving at Europort in Holland, quite a few people arrived from Head Office to test how long it would take to discharge the cargo. It took approximately 15,000 tons per hour, so was done in over 11 hours. The only thing that delayed the ship sailing was missing the tide.

The main problem with these VLCC was the monotonous run they were on; there wasn't much variety in ports.

We did two trips round the Gulf and back, and then it was home on leave.

Not long after getting home, the company was on the phone about attending a Chief Stewards' Course in Liverpool, at the Nautical Catering College. The Board of Trade had introduced a certificate that all Chief Stewards must have, and it included a first aid course, based on "The Ships Medical Guide". The course would run for six weeks and was approved by the Merchant Navy Training Board. I was probably sent because I had done enough time as Chief Cook and was ready to be promoted to Chief Steward.

There were about 20 of us attending the course, from different shipping companies, and we all stayed at the Merchant Navy Hotel in Liverpool.

I had made arrangements while I was in Liverpool to attend the same college and sit a course for Higher Grade Cookery for Ships' Cooks, so I did the course for the Chief Steward, went home for Christmas, and then returned to Liverpool after the New Year to do the cookery course.

While there, a couple of the lads and I used to go to either watch Liverpool FC or Everton FC, whichever club was at home, or we went home for the weekend. After finishing the Higher Grade Cookery Course and passing the exam in February 1971, I went home to finish the rest of my leave.

CHAPTER 10

Near the end of February, the company offered me the rank of Catering Officer for the ship M.V. *British Fern*, which was one of the Tree Boat Class.

I was a bit apprehensive, as it was a big jump from Chief Cook to Catering Officer – the rank had also changed names. The responsibilities of the job also changed to Head of the Catering Department, and you would be in charge of ordering the fresh, dry and bonded stores, and issuing bonded stores on a weekly rota at sea.

The job also involved dealing with Customs, Ship Chandlers, Port Health Authorities, ensuring laundry was put ashore when in port, and handling minor medical problems that arose during the voyage.

Anyway, I decided to accept the rank; I thought, if I didn't, I might not be approached again for a couple of years. As it happened I had purchased a Catering Officer's uniform when I attended the Chief Stewards' Course in Liverpool. Being a sea port, it was the ideal place to be fitted out in the appropriate uniform.

So the Officers and crew flew out to join the *British Fern* out at Aden. It wasn't a full crew change, only the Captain and his wife, who was going out with him, and nearly a full Catering Staff, Cook, 2nd/Cook Baker, 2nd/Steward, and two Asst/Stewards.

During the late 60s into the 70s, the company started allowing senior staff to take their wives with them – Captains, Chief Engineers, Chief Officers and 2nd/Engineers.

Before we joined the ship, we were put in a hotel for the night. But we received a real shock the next day when we learned that one of the

Assistant Stewards had slashed his wrists in his room during the night. They saved his life and he was sent home, but this unfortunately meant there was one staff member short before we had even sailed. This put an extra burden on the other Asst/Stewards, as there was insufficient time at Aden to send another replacement prior to sailing. I wondered what I had let myself in for.

To this day, I still don't know why the young Assistant Steward tried to attempt suicide. Maybe he had problems at home prior to coming out to join the ship, or he had mental health problems. Perhaps something happened the prior evening. I just don't know. The Captain didn't confide in me why, and none of the other Catering Staff offered any information. Luckily he survived the ordeal, thank God.

The ship was loading at Aden to take cargo to Kwinana in Western Australia. Only part of the cargo was being unloaded, but we back loaded in Fremantle, after which we went to another port in Western Australia, called Geraldton.

We then went round the coast of Australia to Westernport, near Melbourne, where we were to top up 20 weeks' fresh, dry and bonded provisions. I was so green, as it was my first voyage as Catering Officer! I should have had the stores sent by barge and winched onboard, straight down the Aft Stores hatch. But they came by truck and had to be manhandled onboard ship, which took nearly all day.

By the time I got ashore – I had said I would see the Cook and 2nd/Cook for a drink later on – it was eight o'clock at night and the pubs closed at ten. The 2nd/Steward had been chatting up a local woman and buying her drinks, but she gave him the bum's rush and scarpered. We left at closing and went looking for somewhere to get something to eat, when the 2nd/Steward smashed a shop window. To cut a long story short, we all ended up in the local police station. He eventually owned up and they let the rest of us off.

I was imagining all the Catering Staff locked up for the night and no breakfast for the Officers and crew next morning.

I had to let the Captain know when we got back. He wasn't too pleased, but neither was the 2nd/Steward when he got the bill for smashing the window. I thought the voyage couldn't get any worse.

After discharging part of the cargo at Westernport and then back loading, we were to go to New Zealand, to Auckland – the second largest city of the two islands. This being the first and last time I visited the country, it left a lasting impression on me. If I wanted to immigrate to a country, this would be it, as it had the same standard of living as the United Kingdom at the time.

The further south you got, it became more rural, with the countryside more like the UK.

I don't know what it was with New Zealand women, but one or two of the crew managed to bring them onboard ship for the passage from Auckland down to Dunedin.

I remember one fireman, whose job was in the engine room as this wasn't a GP-manned ship, had a woman onboard and hadn't been doing his watches on the passage down the coast. So the Chief Engineer went down to his cabin to talk about him commencing his duties. The fireman replied, 'What do you want me to do when I've got my chick onboard?' Needless to say, he was fined several days' pay and entered into the Ship's Log Book. Alas for some, all good things come to an end, and we left the Kiwi Coast and were to proceed back up the Persian Gulf.

I had heard stories about the Great Australian Bight, where on some occasions you could have some atrocious rough seas. On this occasion, we hit some. Apparently there are different tides that have this effect, but the weather was so bad that the ship's main engine started moving from the holding-down bolts.

As the weather calmed down, Head Office was notified and we limped into Fremantle Harbour for repairs. The ship was there for two weeks to sort out the problems, which gave us time to see in-depth the sights of Western Australia. Normally there would be less than two days in port, if you were lucky.

After Fremantle, it was back up to the Persian Gulf to load the next cargo. We loaded the ship at Bandar Mahshahr and were to proceed round to the Lebanon, to the Port of Sydon. At the time, a conflict was brewing between the Arabs and the Jews. The Lebanese who were onboard the ship while we were discharging part of the cargo, appeared to be on edge about a war starting with the Jews.

Not long after leaving, it all erupted in the Middle East and the rest is history.

The remainder of our cargo was to be discharged in Athens, Greece. Before arriving, the company had requested a stores order to top up to 20 weeks' stores of fresh, dry and bonded provision, as the ship was bound for the Isle of Grain in Kent. These were sent by mail. After leaving Greece, the ship was to load the next cargo in Italy, where another one of the Assistant Stewards left the ship with alcoholic poisoning.

So that meant we were now two members of staff down. I would be glad to get back to the UK to get back up to a full Catering Staff.

Before arriving back in the UK, the accounts would be closed. With the majority of the crew leaving, I had to close the bonded account and ensure the figures totalled up correctly.

On arriving at the Isle of Grain, I found out I was about £5 deficient in my bonded account. The Catering Superintendent told me not to worry, some Catering Officers had been hundreds of pounds deficient.

After discharging the cargo, the stores were loaded onboard; the company employed riggers in the UK to load them. It was nice to get back up to a full Catering Staff, and the ship went round to the London Graving Dock to ensure the engine was fully functioning after crossing the Australian Bight.

From London, we had orders to proceed back through the Suez Canal, and round to Bandar Mahshahr.

As the ship was passing through the Straits of Gibraltar, entering the Mediterranean Sea, I had been in the bar onboard when I decided to turn in about 11pm. For some unknown reason, I popped out on deck and all the ship's lights were on, but I never gave it another thought and went to bed.

Next morning I went down to the galley to give the daily dry provisions for the day and went into the Mess Room to give the Steward a shout. The Mess Room hadn't been cleaned and laid out for breakfast, which was unusual, so I went back into the galley to ask the Cook where the Mess Room Steward was. He thought I knew. Apparently late the previous evening, he had jumped overboard – hence all the lights being on. The ship's watch keepers had been searching for about 10hrs looking for the body during the night, but he hadn't been found. He was from the Glasgow area of Scotland.

I thought once we had the full Catering Staff back, things couldn't have got any worse, but there was another dreadful incident.

I had to go down to the Assistant Steward's cabin and pack his belongings, ready to be put ashore and sent home at the next available opportunity.

As the voyage progressed round to Bandar Mahshahr, there were about eight staff that would be leaving there, including me.

We flew to Tehran, the capital of Iran, and stayed in a hotel for the night, ready to fly out next morning to Heathrow. For some unknown reason, the luggage was sent to the airport and we were to get it when we arrived there.

Next day at the airport, everyone got their luggage except me. I was running round like a lunatic trying to find out where my luggage was, being sent from one information desk to another with no luck. The flight was called for people to board and I thought I was going to be left behind, but thankfully my luggage was found, and I was the last one to board the plane. I had certainly been thrown in at the deep end as a Catering Officer, and could probably have written a book about the incidents on that first voyage alone.

CHAPTER 11

I was glad to get home for a well-earned rest, but was informed that I was to attend a Merchant Navy Fire Fighting Course in Plymouth.

It was a course approved by the Department of Trade and Industry, and lasted for four days. It was very intense, and we were put through situations that could easily occur. In fact, it was frightening coping with the scenarios, dealing with the intense heat in smoke-filled rooms trying to get dummy bodies out. You can understand the pressures the Fire Service are put through in their daily job.

After the fire fighting course, it was back home to finish my leave.

The next ship I was to join was the *British Ambassador* in Philadelphia, so I was sent a form to apply for a visa. I returned the form but I had completed it incorrectly, so on my way down to Heathrow I had to make a detour to the American Embassy and explain the cock-up. I thought I was going to miss the flight, but they granted the visa and I made the flight out to Philadelphia.

We had a couple of days in a hotel, prior to joining the ship. The S.S. *British Ambassador* was one of six ships built, with D.W. tons of about 42,000 with midship and aft accommodation, unlike the later ships being built with all aft accommodation. After storing the ship and discharging, we were to go across the Atlantic, round the Cape of Good Hope and load a cargo in the Persian Gulf.

On arriving at the Persian Gulf, we were to load a cargo at Kharg Island, to take down the coast of Australia to Kwinana.

After leaving Australia, the ship had loaded fresh provisions and was to proceed back to the Gulf to a port in Iraq, Khor Al Amaya, which was what was called a sea island jetty. It was a structure off the coast of Iraq, where tankers would go alongside and load cargoes of crude oil.

After leaving Iraq we went back round the Cape of Good Hope, where the vessel slowed down to take on provisions, mail and change the films, then proceeded back to Philadelphia.

On the way over the Atlantic, the ship stopped at the Portuguese Cape Verde islands for bunkers. This voyage would be the longest time at sea, from leaving Australia to arriving in the USA, without getting ashore – two months in total.

By the time we arrived in Philadelphia, discharged the cargo and loaded the ship provisions, there wasn't enough time to even get ashore due to the fast turnaround.

We left the US coast and went to Nigeria to load the next cargo, at the Port of Bonny.

The problem going to Nigeria was that Customs used to board the ship, making all kinds of demands for cigarettes, spirits, cases of lager, soft drinks, etc. The company knew the problem there and waived it aside, as it was cheaper to let the Customs have the ship stores than to delay the ship.

After Nigeria, the next discharge port was Le Havre in France, where I was being relieved of my duties along with one or two of the staff.

Back home I had a couple of months unwinding and adapting to being ashore and getting into a routine. The house I have lived in for the last 66 years has quite a large garden. My father always grew his own vegetables, which was the purpose of these council houses just after the First World War.

But the garden was my dad's pride and joy and I wasn't allowed onto the veg patch till his later years.

The next BP tanker I was to join was the S.S. *British Grenadier*, another one of the Regiment Class. I was to fly out in November 1972 to Ras Al Khaimah in Dubai. This part of the world appeared to be quite popular for joining and leaving ships, with frequent flights from the UK to and from Gulf destinations.

It also gave you the time to hand over the bonded store and provisions accounts to the staff being relieved, i.e. the Catering Officer. After loading the cargo, we were to proceed to North West Europe. The Catering Officer I was relieving had mentioned that he and the Captain didn't get on. He had said he was a dour type of bloke. I was called into the Captain's cabin to sign on, and he said to me that he supposed I had been informed what he was like, but I replied that I would form my own opinion. Welcome to John R. Scott; Scott by name, Scot by nature.

It was round the Cape we sailed on this part of the voyage, and up to Finnart in Scotland to discharge the cargo.

While we were at Finnart, the ship's articles were changed and everyone had to sign off the ship and re-sign, apart from the ones who were going home on leave.

When we were there, I had a letter from home saying my youngest brother had been taken ill. In his younger years he had suffered from rheumatic fever of the heart; now he'd had a flu-like fever, but it was apparently nothing to worry about.

After leaving Finnart, we went back round to the Gulf via the Cape. On the passage round the Cape, the Captain sent for me. He told me to tell the Assistant Steward, who had served him at the evening meal, to get a shower because he smelled. So I delegated the job to the 2nd/ Steward, to tell the lad to get himself smartened up.

The Captain could be abrupt at times. You never knew what to expect from him, whether it was a dressing down, or he wanted a bottle of gin or whisky out of the bond locker.

After arriving in the Gulf, we were to load a cargo and go to Genoa in Italy. This involved going back round the Cape because this class of tanker was too big to go through the Suez Canal.

When we got to Genoa, the mail I received from home told me that my brother's health was good and he had recovered fully.

From Genoa we were to go back round the Cape. But four days after leaving port, we got orders to go to Port Harcourt in Nigeria to load a cargo to take to Curacao in Netherlands Antilles.

After arriving at Port Harcourt in Nigeria, the usual mob arrived onboard ship; Customs and Port Officials. I have to give credit to the Captain on this occasion. It must have been the first time the Port Officials left the ship with no backhanders. Although they tried their best in the slop chest locker, I didn't have the authority to let them have anything without the Captain's permission, so they went ashore with their tails between their legs.

Halfway across the Atlantic to Curacao, I was notified that I was to be relieved when we arrived there.

I remember one night I was starting to get ready to close the bond and provisions accounts, when a Deck Cadet knocked on my door saying the Captain wanted to see me. I trooped up to his cabin, knocked on his door and went in. He told me to sit down and asked if I wanted a drink; it wasn't like him. Then he dropped the bombshell that my brother had passed away. The last mail I had received in Italy had said he was fine, but apparently a week or so later he had suffered a relapse and passed away in Hull General Hospital.

I was glad I had enough work to occupy my mind and not dwell on the sadness until I arrived home.

On arriving home, unfortunately I had missed the funeral service and burial, but at least I was home on leave to grieve. The good thing about BP as a company was they had a very good personnel back-up team, who managed to get staff home at distressful times.

CHAPTER 12

On a few occasions while home on leave, you could be sent to staff conferences where Shoreside Management and Seagoing Staff would be together to discuss the future structure of the company and where it was going. I was also sent on a Personnel Relations Course down in Brighton, with outside instructors teaching us how to get the best out of staff at sea in their daily working practices.

After the Staff Conference and the Personnel Relations Course, I went back home to finish my leave. The next ship I was instructed to join was the M.V. *British Gull*, one of the bird boat class, out in Singapore.

The ship was about to go into Dry Dock out there, but the ship's personnel who were going out joined before we actually went into Dry Dock. At the time a friend of mine, who I had gone to school with, was living out in Singapore. We'd exchanged letters and he had said to look him up if we ever called there. What a coincidence. He was out there working for Parsons, installing turbine engines for a power station.

The *British Gull*, the ship I was joining, was staffed by BP Officers, but was manned by an Indian crew from the sub-continent; my first ship with an Indian crew.

Apparently when India was granted independence, some agreement had been made between the Indian Government and the British Government, that British shipping companies would take a number of Indian crew on their ships.

I don't know the exact number the BP Tanker Company agreed to take on their ships, but when there were Indian crews, the ships had to be adapted to suit their different ablution and toilet facilities.

The Indian Catering Department was normally from Goa – an ex-Portuguese Colony – with the majority being of Christian Religion. They could handle pork and beef and other foods that normally Hindus and Muslims wouldn't, due to their religions.

After joining the Br *Gull*, we went into Dry Dock for approximately two weeks of repairs. Once we'd settled in on the ship for a few days, I decided to pay a surprise visit to my mate. I ordered a taxi to take me to where I hoped he was staying, and asked the taxi driver to wait. I knocked on the door and when he opened it, he couldn't believe it. He nearly dropped through the floor. We had a couple of nights out while I was in Singapore, as he had his wife out there with him. He eventually emigrated with all his family – mother, father, wife and two brothers – to Western Australia, where he still lives.

After leaving Singapore, we went up the Persian Gulf to load a cargo for Australia. It was one of those cargoes that involved different types of fuels, so it looked like we would be on the Aussie Coast for a while. The first port was Kwinana, to discharge so much of the cargo, then back load to go round to Adelaide. After Adelaide, it was on to Melbourne. We then took on more cargo and were to proceed to Sydney. Going into Sydney, we passed the magnificent Opera House and went under the famous Sydney Harbour Bridge, which apparently had been shipped out from the Dorman Longs Shipyard on the Tyne, and is the exact replica of the one built on the River Tyne.

After leaving Sydney, we went up the Aussie Coast on the eastern side to Brisbane. After the number of times I have been to Australia, I must have seen more of the Continent than most Aussies.

Leaving Brisbane, we went to Fiji, to the capital Suva. We then went to Port Moresby, which is in Papua New Guinea, to unload the remainder of the cargo. This was another part of the world where headhunters used to prevail. When we eventually left the Aussie Coast, it was back up the Persian Gulf, but we went an unusual way, through several islands. It was strange to see young children running along island beaches as the ship passed through the Straits of Sumatra.

After arriving in the Gulf, the next cargo we were to load was in Iran, bound for a port in India. On arriving at Kandla in the Gulf of Kutch, another ship was tied up on buoys so we were told to anchor. The Captain dropped one of the anchors, but the tide was flowing so fast he decided to drop the other one, which was fatal, as the anchor cables were becoming entwined. When the other ship left the buoys, a tug had to come out and unravel the anchor cable before we could berth.

I think the Captain was glad to get out of that port as it had been his worst nightmare.

We went back up the Persian Gulf to Iran to load another cargo. One of my problems on an Indian crew ship was filling in the crew's personal manifest for the local Customs when entering port. It used to take about two or three hours getting them all organised, as there were about 35 crew members to ask about 20 different items on the manifest. Some couldn't write their names, so I would have to get them to put their thumb print as a signature.

After loading the cargo at Bandar Mahshahr, we were to go back to India to discharge the cargo at Bombay. I was told on the way down that I would be leaving the ship at Bombay and flying home from there. I left the *British Gull* on the 6th December, 1973, and went home for Christmas.

Not long after the festive season, the company hierarchy informed me that I was to attend a Ships' Captains' Medical Training Certificate in Liverpool, approved by the Department of Trade and Industry, and which had an examination at the conclusion of the course.

It was based on the book, "The Ships' Captains' Medical Guide", which was the guide used by Catering Officers/Chief Stewards to remedy minor medical incidents at sea. Luckily I had never had any serious accidents to deal with throughout my seagoing duties as Catering Officer.

My next ship was the S.S. *British Cavalier*, which was one of the Regiment Class, 52,000 dead weight tons. It was out in Singapore being fitted out for an Asian crew.

We flew from London to Orly Airport, to join a French Cambodian aircraft which was flying out to Phnom Penh. One or two of the passengers weren't too happy at the time at being so near to the Cambodian border, with the Vietnam War not too far away.

On arrival at Phnom Penh, there was a total shutdown at the airfield, apart from locals leaving the plane. The plane refuelled and we were soon on our way to Singapore, much to everyone's relief. Unfortunately on this occasion my old school mate, who had been working there, had completed his two year contract and returned home to the UK.

After leaving Singapore, the ship went back up the Persian Gulf to Kharg Island, which belonged to Iran and was a crude oil depot for larger tankers.

Apparently Kharg Island came into production due to the Abadan Refinery, up the Shatel-arab River, being run down, and Iraq and Iran were at loggerheads about who should dredge the river. Bandar Mahshahr had come into full production as the main refinery in Iran.

After loading at Kharg Island, the ship was to proceed to Japan to discharge the cargo at a port called Mizushima.

Being on the Indian crew ship, the diet was limited. They would have MK Mutton five days a week, fish on Friday – mackerel or herring – and Sunday was chicken; all served with rice, cooked for them by an Indian cook.

Before arriving at Mizushima, the company had notified the ship to top up to 12 weeks' stores. With Indian crew stores, the mutton and chicken had to be slaughtered following the Muslim religion, to bleed from the neck (MK). Certificates also had to be produced or the crew wouldn't accept it.

On arriving in Japan, the stores were shipped and, after dealing with the Port Authorities, Customs and Ship Chandlers, I made sure the MK Certificates were produced. They were stamped with the Muslim Authentic Stamps.

Not long after leaving Japan, the Serang, who was the head man on the Indian crew, went up to see the Captain and said the chickens that had been shipped in Japan weren't MK. The Captain sent for me to show the Serang and his delegation the MK Certificates, which were genuine, but they would not accept the authenticity of the paperwork.

The only reason I could think of at the time was that, because the chickens still had the heads and feet on, this would be calculated in the amount of weight in their weekly ration.

About a week later we had to pick up some spare parts, passing Singapore, so we took on more MK chickens at the same time.

After leaving The Straits of Singapore we were to go back up the Gulf, and the Captain was leaving the ship on arrival at the loading port. The Captain who was leaving was a Scot, and it was a joke between him and me that his replacement would be my favourite Scotsman, J.R. Scott. About four hours before he was to be relieved, he came down to my room to let me know that the man replacing him was indeed J.R. Scott. At least I knew what I would be letting myself in for.

After loading at Kharg Island, we were to take the cargo to a port in Madagascar. It wasn't a favourable place to take any stores, so it was by-passed.

After leaving Madagascar we went back up the Gulf to a port in Iraq, which was a sea island jetty, miles from anywhere. There wasn't much chance of getting any fresh salads or vegetables there, even though our stocks were practically depleted.

From there we were to go to Kwinana in Western Australia to discharge the cargo. On the way down, I was getting a little worried because certain types of food stocks were a bit short. I remember going round on the weekly inspection and the Captain asking where everything was when we visited the meat room. The following week, when we went round, he said, 'There's no use wasting your time going in the meat room, as there's f--- all in there.'

I've never been so glad in my seagoing career to reach a port; I think, to mention one item, we only had about 2lbs butter left. I nearly handed my notice in next time on leave.

We stocked up with stores in Australia, as they were favourable in price, then went back up the Persian Gulf. Halfway through the passage, I was told I was to be relieved at the next loading port, which was to be Bandar Mahshahr. I left the Br *Cavalier* on 23rd June, 1974; the only member leaving. I flew to Abadan, stayed the night there, then flew up to Tehran and caught a connecting flight to Heathrow Airport. At least there were no luggage incidents that time!

CHAPTER 13

After being home for two months, you were often glad to get back to sea, although I had some fantastic holidays when I was on leave. I went twice to the Seychelles, visited the carnival in Rio, and travelled to three different islands in the Caribbean.

The next voyage I was to join another of the Regiment Class, S.S. *British Lancer*, at Europort in the Netherlands. On this occasion, I was going out to relieve the Catering Officer while he went home on leave.

What used to happen was a Catering Officer, Captain Chief Engineer, etc, would be appointed to a certain ship for two years, but he would be relieved to go home on leave and then rejoin the ship after his allocated time off.

After leaving Europort, we went down the West African Coast to one of the Nigerian ports, called Port Harcourt. We were to load a cargo and go back to Europort to discharge it; about twenty days' round trip.

Back in Europort, after discharging the cargo, we were to go round the Cape of Africa, and then up the Persian Gulf to load the next cargo. We went to Kuwait, to a port named Mina Al Ahmadi, where we were to load the next cargo and head to Singapore. On the way down through The Straits, I was told that I would be relieved at Singapore as the permanent Catering Officer was due back from his leave.

The trouble when you were relieving other staff was that you might only get a few days at home before you were sent back to relieve someone else. Fortunately, I was lucky enough to get ten days at home, before I was sent out to join another ship – the M.V. *British Commerce* out in the Gulf at Ras Al Khaimah, which is part of the United Arab Emirates.

After joining the *British Commerce*, there were a couple of days for a handover before arrival at the loading port, which was Ras Tanura in Saudia Arabia. The crew who were on board had been on the ship for five months, after signing on at Finnart in Scotland. They weren't the company's staff, but had been appointed from the Shipping Federation Pool.

After loading at Ras Tanura, we were bound for a port in Thailand. The ship was on a six month charter to Shell Oil Company, and when I joined it was halfway through the charter trading from Ras Tanura and the Port of Sattahip in Thailand. I have never been on a ship which had so many breakdowns, apart from the *British Trust*. Every other day there were problems with the engine. It was a nine cylinder Sulzer engine and I think in the two months I was on there, every cylinder was changed. I felt for the engineers trying to keep the engine running.

I have never known the morale so low. There were about 60 Officers and crew, and the crew didn't know when they would get home as they were not on contract to BP.

On arrival at the Port of Sattahip, I was forewarned about the Customs Officers who boarded the ship. They took 20,000 cigarettes, a couple of cases of whisky and anything else they could get their hands on. It was as bad as the Nigerian ports for the corruption, and happened every time the ship went to Thailand.

After leaving Sattahip, it was back up the Gulf to load again at Ras Tanura. When the ship was running, it used to do a really good speed, up to 18-19 knots at top speed, but unfortunately it wasn't consistent.

With it being a pool crew from the Shipping Federation, the Captain limited the crew's bar to 8 x 24 cans beer daily; 2 x 24 at lunchtime and 6 x 24 in the evening. I used to put it in the fridge to cool it down, as the beer locker was quite warm. The evening issue was picked up about six o'clock at night, and would be drunk by nine o'clock. The trouble with it being rationed was there was a mad rush for everyone to get their share.

After loading the ship at Ras Tanura, we were to discharge the next cargo in Japan. On the way down, the company decided to employ

Sulzer engineers, who were to join the ship at Singapore on the way to Japan, and then leave at Singapore on the way back. When we left Japan, the Sulzer engineers thought they were being taken off the ship at Singapore, but the Chief Engineer decided they were needed for more engine duties so he kept them till the next loading port. They weren't too happy.

After loading again at Ras Tanura, it was back down to Sattahip, with the usual Customs' demands. Prior to leaving Sattahip, everyone had to appear before the Port Authorities to ensure we were all on board before sailing, and to give their signature. Some of the crew that had been ashore were drunk and I don't know how they managed to sign their names.

We sailed back up the Persian Gulf and, on the way up, I was told I was being relieved at the discharge port, which would be Jebel Dhanna.

Ras Al Khaimah was normally the port where BP staff were relieved. Staff would fly out to Dubai Airport, then be taxied down to Ras Al Khaimah and board a launch, which would normally have salads, fresh veg and sometimes a laundry exchange to be taken out to the ship, as it was sailing up to the loading port.

On this occasion the ship wasn't notified that the 4th Engineer, who should have been joining the ship, hadn't turned up. So the original 4th Engineer left the ship with no relief to take over.

The Chief Engineer wasn't too pleased at being an engineer short. It couldn't have happened to a nicer bloke – he was an unpopular man.

The port of Jebel Dhanna was in the United Arab Emirates. I left the ship about 8 o'clock at night and was taken to an office ashore. I was the only one leaving the ship there.

The agent said a taxi would pick me up about 10 o'clock and take me across the desert, which would take about six hours. I thought he was joking; little did I know!

I still don't know how taxi driver found his way across that desert, but about 5 o'clock in the morning we got onto tarmac road. Then I was taken to Abu Dhabi, to a hotel for the day. I was so glad to get off the

British Commerce, it had been a nightmare.

After my day in the hotel in Abu Dhabi, I was put on a flight back to Heathrow Airport, for a couple of months' leave.

On this occasion I had decided to have a couple of weeks' holiday in the Canary Islands, so I chose the island of Tenerife. A few days before I was to fly out there, a Dutch KLM Aircraft was flying into land when the aircraft tragically crash-landed on the tarmac due to fog, if I remember. I phoned the travel agents to enquire about the holiday I had booked, and ended up phoning every day until eventually flights were allowed onto the island.

But it was distressing when we did fly onto the runway at Tenerife, as debris was still lying around on the tarmac as we landed.

The next ship I was to join was the M.V. *British Security*, at the Grangemouth Refinery in Scotland, on 13th April, 1975.

The ship was loading to go to Belfast in Northern Ireland, and we were to go round the top of Scotland, via the Pentland Firth. After leaving Grangemouth, we sailed under the Forth Railway Bridge and round the Irish Sea into Belfast, where the *Titanic* was built at Harland and Wolff Shipyard.

After leaving Belfast, we were scheduled to go to Dublin in the South of Ireland, to discharge part of the cargo, with the remainder to be discharged at the port of Cork.

Not far from Cork, there was a huge sea port at Bantry Bay, which was built to take these VLCCs – "Very Large Crude Carriers" – over 200,000 tons. Bantry Bay was an ideal place, with deep water facilities for ships to berth there.

On the *British Security*, I was relieving the Cat/Officer while he was on leave.

After Cork, we went round to the Isle of Grain in Kent to load a cargo to take to Cagliari in Sardinia.

Not long into the Mediterranean, we were having the weekly inspection of the ship's accommodation, when there was a huge bang and the ship started shuddering. The Chief Engineer said we had hit something. Apparently, one of the blades of the propeller had snapped off, and the balance, being only three blades, was causing the ship to shudder.

After notifying Head Office, we limped into Cagliari to discharge the cargo.

After the discharge we went out to a safe anchorage, and the company notified us that the ship's staff would have to saw the opposite blade off the propeller, to balance the two remaining blades.

After about three days and cutting about six inches through the blade, the company decided to send a welder out to burn the blade off. He did it in less than half a day, and the part that had been severed off was kept because it was bronze and, therefore, expensive.

We loaded a cargo in Sicily for North West Europe. Sometimes you didn't know where the cargo was to be discharged until you were on your way. The ship's engine appeared to be coping well with just two propeller blades; normally the speed would be approx 14 knots with a four bladed propeller, but with two it was about 10 to 11 knots. On the way, we got the orders to discharge the cargo at Rotterdam, and then to go to London Graving Dock to have a new propeller put onto the shaft. I was to leave the Br *Security* at the London Dock. As the next ship was also a relief job, I managed to get a couple of weeks' leave before I was to join the *British Fidelity* at Hamburg. This would be the fourth "ity" boat I would be joining.

The Br *Fidelity* was trading round North West Europe and the UK. Before I joined it, ship's stores had been taken on at Hamburg, and then we had orders to take a cargo to the Isle of Grain in Kent. Normally on arrival, the ship would have been visited by the Shore Catering Superintendent. He came up to my office after he had been down at the fridges, and asked me where we had taken stores last – I told him Hamburg.

He told me that if the Board of Trade had visited the ship, they would have condemned the beef that had been shipped there, as it was inferior quality.

Apparently "Irish Intervention Beef" was being exported to the then Communist Bloc countries and then imported to Germany. The Ship Chandlers in Germany shouldn't have sent this beef, being inferior quality, and should have stuck to the agreement they had with BP Tankers about the types of beef they should have supplied – from Argentina, Australia, New Zealand or Uruguay.

After leaving the Isle of Grain, the ship went to Rotterdam to load then discharge the cargo at Antwerp in Belgium.

One of the disadvantages of being on the coast was that no sooner had you filled the Customs Personnel Manifests than you would be doing one for the next country, as Holland, Belgium and Germany were so close to each other.

The good thing, though, about coastal trading was that the time appeared to go quicker, as you were in and out of port sometimes every other day.

After being on the Fidelity for just over two-and-a-half months, I was relieved of my duties at Rotterdam and went home on leave on the 11th September, 1975. This time – I don't know why – I was home for nearly four months. I went on a Shipboard Management Course for five days, but I should have been back at sea in December. It was unusual to have nearly a month's extra leave, and I even got Christmas at home so I wasn't complaining. Eventually I got a letter to join my next ship, which was the *British Neath* – one of the River Class.

CHAPTER 14

There were about 12 staff joining in Egypt, in the Bay of Suez on the Red Sea side of the Canal.

After leaving Suez, we went to Aden in the Yemen and loaded a cargo to take to Durban in South Africa.

After Durban, we went back up the Gulf and loaded at Bandar Mahshahr. At about this time, the Shah of Iran had been deposed and the Islamic rulers had taken over in Iran; Ayatolla Khomeini and his followers. Instead of the Gulf being referred to as the Persian Gulf, it was now the Arabian Gulf. I hadn't realised at the time, but I was notified at a later date that I was to be appointed to the M.V. *British Neath* for two years. After loading at Bandar Mahshahr, the ship was to proceed to Kwinana in Western Australia and unload some of the cargo, and reload to go on to the Aussie Coast.

After Kwinana, we sailed round to Adelaide, having topped up with fresh dry and bonded provisions.

When we arrived at Adelaide, it was a Sunday. The CPO had asked me if I was going ashore there, but I said there weren't any pubs open on a Sunday. He told me there was a Working Men's Club in Adelaide, which was affiliated to the "Club and Institute Union" (CIU) in the United Kingdom, and he had his club card with him. So we went ashore, they let us in and we had a few pints and a game of snooker.

After Adelaide, we went round to Sydney and discharged the remainder of the cargo, then travelled back up the Arabian Gulf to load a cargo for Singapore.

While I was on the *British Neath*, there must have been some agreement with the National Iranian Oil Company and BP Tanker Company that Iran would take over a number of ships but they would be still owned by BP, registered in London and manned by BP personnel.

So after loading in Iran, the ship sailed to Singapore to discharge the cargo and had its name changed from the M.V. *British Neath* to M.V. *Mokran*.

I was leaving the ship in Singapore but would be rejoining at a later date, probably out in the Arabian Gulf.

I left the Br *Neath* on the 29.05.76 at Singapore. After my statutory leave, I was to rejoin the M.V. *Mokran*, ex Br *Neath*, out at Dubai.

I rejoined the ship on 14th August, 1976. With the Iranians running it, the ship was mainly trading between Bandar Mahshahr and Bandar Abbas or Kharg Island, mostly in the Arabian Gulf. The Port of Register for M.V. *Mokran* was at Khorramshahr in Iran.

You had to be careful while trading in Iranian territorial waters when the Customs came on board. Although we had bars on board ship, on the *Mokran* it was frowned on to exhibit or display alcohol. Occasionally the Customs would want to buy the ship's whisky, which was forbidden ashore, and it was illegal to sell to locals from the ship's bond locker. Sometimes the Customs would try and exert pressure on you to sell whisky to them. I don't think I would have liked to spend time in an Iranian jail for the sake of a bottle of whisky.

Occasionally the ship would go out of the Gulf with a cargo to Singapore, Australia, or South Africa, and top up with fresh, dry and bonded stores. Stores in the Gulf could be quite expensive locally, as the majority of goods had to be imported. On the M.V. *Mokran*, the Captain was from Wales. He was a decent bloke and when he went on leave, the Chief Officer would take over as Captain; he had his extra Masters Certificate, and was competent to take over the responsibilities of being sea-going Captain. In the meantime, another Chief Officer would be sent out while the permanent Captain had his leave. When he returned, the promoted Chief Officer would go on leave, and return as Chief Officer. The term was YoYo Promotion, as with the Chief Officer, then Captain.

On this occasion, we had Christmas on the *Mokran* in the Gulf.

Just after Christmas, I was told I would be relieved in the middle of January; in fact it was the 19th January when I left the *Mokran* to take my leave. While I was at home on leave, I was having a bit of a problem with my right leg, which had swollen up at the knee joint. I wasn't getting much satisfaction from the local doctor, so I paid to see a specialist. After about three weeks of physiotherapy and an X-ray, he came to the conclusion he didn't know what was wrong.

So after an extended leave, with a month on the sick, I rejoined the M.V. *Mokran*, once again at Ras Al Khaimah in Dubai.

The Catering Officer who had relieved me told me he had thought I wasn't coming back, until I explained about the hospital treatment I had during my leave. With the ship trading mainly in the Gulf area, it was a good job we had air-conditioning, as the temperatures there in summer were really high. Halfway through the tour of duty, the Captain had to change the ship's articles, as too many staff were on board. New articles had to be opened, so we all signed off, then re-signed on.

Sometimes we would load a certain cargo and the ship would anchor off at Khasab Bay, which was a safe anchorage. Small coastal ships would come alongside our ship and we would pump some cargo into them, and they would take it up one of the coastal creeks to discharge. It meant we could be there for three or four weeks at a time before the ship was empty.

I was notified that I was to be relieved of my duties and it would be my last tour of duty on the M.V. *Mokran*. I left the ship on the 2nd October 1977, at Bandar Mahshahr, and flew up to Tehran to catch a connecting flight to Heathrow.

After my statutory leave, I was told that I was to join the M.V. *British Trent*, one of the river class, at Ras Tanura in Saudi Arabia. From there we went through the Suez Canal and round to Rotterdam to discharge the cargo. We then went to the Isle of Grain to store the ship with fresh, dry and bonded provisions. The good thing about storing in the United Kingdom was that riggers were there to do the work, instead of the crew or catering staff having to do it.

I was only on the ship to allow the permanent Catering Officer to have his leave, and at the time we were coasting round North West Europe. Eventually I was taken off the ship at Teesport, near Middlesbrough.

Sometimes it was beneficial to be relieving, as you could be at home longer than you would normally. On this occasion I was given three weeks' extra leave, courtesy of the company.

The next ship I was to join was the S.S. *British Trident* at Rotterdam; it was one of the company's VLCCs. The size of these ships was awesome. They were huge when loaded, but when the ship was discharged, it was astonishing to see the vastness of them.

In around the late 70s, quite a lot of shipping companies were having these VLCCs built, but there was a glut of oil, so they quickly went out of fashion. The cost of running them was astronomical, so what BP did was to run them on one boiler at half speed. Instead of doing 14 knots, they ran them at about 8 knots.

The two month round trip from Rotterdam to the Gulf and back would then take about four months and be economical to run.

When I joined, the ship was going to Nigeria to load a cargo, which would normally take about three weeks' round trip, but on this occasion took us just short of seven weeks. We arrived back at Rotterdam, or Europort, as it was one of the few ports in North West Europe at the time where these VLCCs could discharge their full cargo. Bantry Bay in Southern Ireland was another.

At the same time at Europort there was a full crew change, plus we stored the ship with provisions as it was going round the Cape to load in the Arabian Gulf.

The only thing I disliked about these types of ships was the speed and the monotony of run they went on. When they were doing full speed, which was about 14 knots, it took 8 miles to shut the engine down and fully stop the ship.

When approaching the Cape of Good Hope, stores would have been ordered – mainly fresh salads and vegetables, fresh fruits with a change of films. These would be sent out by helicopter while the ship was at sea, and landed on deck. Normally there would be two helicopters sent out.

After arriving at the loading port, which was Mina Al Ahmadi in Kuwait, it took eleven hours to load the ship once alongside the jetty.

From Kuwait it was back round the Cape to Europort. On the passage back to Europe, I was told that I would be leaving the ship just off Robben Island.

This meant my relief was joining by helicopter, and I would be leaving by helicopter – a first for me.

The stores, films and mail would be coming out with the staff joining off Cape Town, and this was the first time I had handed over on the ship's deck to my relief. I shook hands and was lifted on the helicopter.

We flew from Cape Town to Johannesburg for a connecting flight to Heathrow.

CHAPTER 15

Part of the way through my leave, I was asked if I would attend the company's Storing System which was at Harlow in Essex.

In the 70s, BP decided to set up a Storing System ashore and also on board ships, named the Visi Record System. And the person nominated to record all the items issued from the system would be the Catering Officer.

The Catering Officer's job at the same time was to be downgraded to Cook/Steward, but somebody in Head Office decided to keep the Catering Officer job, and he would be the person delegated to look after the system and be responsible for entering and deleting the items on the Visi Record System. The warehouse where it was set up in Harlow was huge; there were about six sea-going staff there and the majority were shore personnel.

The post would be for about six months, so we had to find digs down there and sort out lodging arrangements. We used to finish at Friday lunchtime and be allowed to travel home for the weekend, but we had to be back by Monday lunchtime to commence the start of the week.

The biggest problem I found with the system was that on the smaller ships there wasn't enough space to centralise the number of sections in the Visi Record System. On the VLCCs "Very Large Crude Carriers", the store rooms were big enough to have the majority of the sections in one large store room.

Another problem at sea was in the paint section, or packing and jointing, as they weren't under control. The paint would be stowed in the forecastle, and the packing and jointing would be in the engine room.

You hadn't a clue what was being used, as the people using the stock never handed the chits in to be booked on the Visi System. The only time you found out what stock you had was when you did your annual inventory to order next year's stock.

The warehouse in Harlow was where the majority of the stock was kept. The twelve month order would be sanctioned by Head Office, then sent to the warehouse where it would be hand-picked into cartons and then shrunk-wrapped onto pallets, ready to be shipped out to the appropriate ship at a convenient opportunity.

Also at the Harlow warehouse, in another part of the store, would be spare parts for certain types of machinery on the different classes of ships, from various countries where BP had purchased them. When the annual stores were sent out to the appropriate ship, they would be stored until we left port and unpacked. There could be anything between 20 to 30 shrunk-wrapped cartons on pallets. On the VLCCs it was ideal, as the majority of the pallets would be put outside the store room, ready to be checked off with the invoices.

On the smaller tankers, everything had to be manhandled and carried to the appropriate store rooms. There were 14 sections on the Visi Record System and each item had its own card. The sections included paint, fire & safety, nuts, bolts, washers, general stores, packing & jointing, cleaning equipment, catering equipment, clothing & safety boots.

Each section varied in numbers, so there could be 3-4,000 items in the nuts, bolts and washers section, or maybe 150 items in the cleaning equipment section. It might take three or four weeks before everything was checked off and then entered and adjusted onto the Visi Reload System.

During the next 12 months, each individual section had to be checked off to keep the system up-to-date. This was okay if you were on the ship for two years, as you had some sort of chance of keeping it up-to-date, but if your relief wasn't too keen, it could develop into a losing battle.

One of the sea-going staff, who was in Harlow at the same time, was from Scotland and he used to fly home every weekend to Edinburgh, then fly back on Mondays.

One weekend I travelled home, I decided to get off the train at Doncaster rather than Wakefield. The weather was atrocious, with heavy snow, so I got a taxi from Doncaster to Hemsworth – a bad mistake. The car driver got me as far as Royd Moor. As the snow had drifted, I had to get out and walk, but a snow plough had cleared the road further down so I walked the last two miles home. It was a good job I hadn't a suitcase with me.

Getting back was a lot easier and, after doing six months at the warehouse, it was home to finish the remainder of my leave.

While I was on leave, on the spur of the moment I decided to have a couple of weeks in the South of France, at Cannes.

It was quite an experience, and from one extreme to another down there with the prices. You could live quite comfortably if you kept out of the bars and cafes on the front of the promenade, and the supermarkets weren't too expensive. I sent the warehouse staff in Harlow a card, as they were a good bunch of lads.

Unfortunately, all good things come to an end and, after I got back home, I was notified to join the S.S. *British Renown* at Rotterdam on 1st August 1979. Since leaving the British Trident off Robben Island, I had been home for nearly 12 months, on leave and in Harlow for training.

The British Renown had been out in the Pacific for the past two years, anchored as a storage tanker off the coast. Ships would come down the West coast of America from Alaska and pump the crude oil into the Br *Renown*, which would discharge into smaller tankers. They would then take the oil through the Panama Canal round to the East Coast of America to be refined. It was a contract which BP had had for the last two years, but the contract was now up.

The ship had sailed round Cape Horn, as it was far too big to transit the Panama Canal. It had been fully loaded from leaving the anchorage in the Pacific, round into the Atlantic and up to the discharge port of Rotterdam.

Once the ship left Rotterdam, it was to go into dry dock in Portugal.

The main reason the ship had to go in dry dock was, having been at anchor for the last two years in the Pacific, there was an accumulation and build-up of barnacles on the ship's hull. When the vessel was at full speed, it would travel at between 14 or 15 knots, but the mass of barnacles on the hull would slow the ship down to about 12 knots.

This would mean a considerable cost in time lost over a period of 24 hours and extra consumption of fuel.

So it was imperative these barnacles were blasted off the hull of the ship. We were in dry dock for about two weeks, with a few other problems to be sorted out. While we were there, the company decided to entertain the shipyard management with a couple of buffet evenings on board ship.

Eventually we left Portugal to go round the Cape. The BP Tanker Company had 14 of these VLCCs, all of them built in various shipyards in Japan, and there were a further six jointly operated through the Iranian Tanker Company.

We arrived off the coast of the Cape of Good Hope where we took stores and mail. At the time, the company was looking to save money on the ships, so the GPs' crew's working week was reduced to Monday to Friday, with the weekend work knocked off. Overtime was stopped, which didn't go down too well. The only hours worked at weekends were the watch keepers, and obviously the Catering Department as they had to provide meals seven days a week.

I was notified that I was to leave when the ship arrived in the Gulf, as I was only relieving on the Br *Renown*.

I left the ship at Ras Al Khaimah on the 6th October, 1979, to await my next appointment.

Around about this time, Iran and Iraq had started their skirmishes against each other, and there was bitterness about who should dredge the waterways between the two countries.

Iran was also having two oil tankers built in Japan, in Yokohama, and BP Officers – with Indian crews aboard – had the contract to run them. They were specially built with not much depth, so they could navigate up the Shatel Arab Waterway to Bandar Mahshahr.

After my leave from the Br *Renown*, I was notified to join the M.T. *Shirvan* in Kuwait, at the port of Mina Al Ahmadi, on 18th December, 1979.

It was mainly the Captain, Chief Officer, Chief Engineer, 2nd Engineer, Radio Officer and myself as Catering Officer, who would be joining to relieve the staff on board. We would be on board while the permanent staff on the *Shirvan* went home on leave, the remainder leaving at intervals at later days. I relieved a bloke called Dave Wooley, who had joined the M.T. *Shirvan* in Yokohama with the rest of the officers and crew.

He had set up a system on board the *Shirvan*, which was later to be incorporated onto BP ships. But he never got the credit for doing so. At the time it was a bit crudely set up, but it was fine tuned by the storing departments in London and Harlow. It was a system where, over a period of time, you could tell the cost of food per person, per day, and the total cost of victualing for the period of the voyage.

At the end of the victualing period, it would tell the Catering Officer if he had over-expended or under-expended on the victualing period of the voyage.

After joining the ship at Kuwait, we mainly traded in the Arabian Gulf between Bandar Mahshahr and Kharg Island. It was getting a bit risky in the Gulf with the situation between the Iranians and Iraqis. Occasionally jets would fly over. Although you didn't know whose they were, they never bombed the oil refineries or crude oil installations. Our Captain was from the North East, and sometimes he was a bit anxious coming out of port at Bandar Mahshahr, going down the river fully loaded.

I remember being on the Bridge and he was having kittens, because there was only a foot between the hull and the sand bank. He thought that if the ship ran aground, he would be brought before the Revolutionary Guards ashore, although there was a pilot who took the ship down the river.

After being on the *Shirvan* for about seven weeks, the permanent staff rejoined and we were sent home for 10 days before joining the sister ship, *Tabriz*, in Japan.

We flew to Anchorage in Canada, then across the Pacific into Tokyo, then onto a connecting flight to Yokohama.

It was mainly senior staff who had been sent out, with the remainder of the officers joining at a later date, and the Asian crew coming out a week prior to leaving Japan.

I must say, the Iranian Shipping Company gave the ship's personnel a very generous allowance to live on for the month we were out there. We stayed in a hotel before joining the ship, and the shipyard management took us for a couple of nights out to a typical Gensia House.

We also went to a fish restaurant, which was an eye-opener. The first two courses were raw fish. I didn't mind the raw tuna, but the raw prawns with the heads and tails on were a bit off-putting. The Captain said, 'You must eat some, or you will lose face.' I whipped the heads and tails off and swallowed them. The remainder of the dishes were cooked, thank God.

I remember one night I went out with the Captain to a Japanese pizza parlour, and it was bouncing down with rain. We had our raincoats and umbrellas but, as soon as we walked into the parlour, you could have heard a pin drop. All the Japanese must have thought we were US Military Police, and they were pointing fingers.

I still don't remember how we managed to order our pizzas, not speaking any Japanese.

One weekend in Yokohama, a few of us had been out on a booze-up as it was getting close to joining the ship to sail.

I got back to the hotel and went to bed. Next thing I remember was my room door swinging open, and there was a naked bloke trying to get into the wardrobe, having a piss. I thought I was having a dream. I shouted to him, and the next thing I was pinned to the bed with the wardrobe on top of me. Then he had gone. I managed to get up and put the wardrobe back, but I didn't know what to do. I reported to the Reception Desk about what had happened, but they weren't too alarmed, as there wasn't much damage done.

When we joined the ship, who should be next to me as the Third Officer but the guy who had urinated in the wardrobe in the hotel.

We started getting the ship ready for going to sea, getting all the equipment and bedding out in the rooms, making up beds in Officers' cabins and distributing the crew's linen. Then we left Yokohama to go up the Arabian Gulf.

M.T. *Tabriz* was named after one of the northern cities of Iran and, along with its sister ship, *Shirvan*, was registered at Khorramshahr. It was owned by the National Iranian Tanker Company in Tehran.

The first cargo was loaded at Bandar Mahshahr and we had to take the oil to Kharg Island to discharge.

The time it took from Bandar Mahshahr to Kharg Island was about one week, round trip. The Customs never came down to the ship at Kharg Island, but one officer always came after we berthed at Bandar Mahshahr. He was forever demanding a bottle of whisky, until I used to mention the Captain, then he would back off. If I had sold him a bottle and he had been caught by the Revolutionary Guard, I would have probably been marched ashore.

After I had been on the *Tabriz* for a couple of months, the company relieved the permanent staff and we went home on leave to rejoin at a later date. While I was on leave, the Iraqis and the Iranians were at it hammer and tongs.

I was sent on a Survival Course at Felixstowe on the South East coast for four days, basically to learn what to do if – God forbid – you had to abandon ship. You were put through the paces in a life-like experience in a dinghy. We also had a day on the beach letting off several different pyrotechnics.

After my leave, I was sent back to rejoin the M.T. *Tabriz* at Ras Al Khaimah in the United Arab Emirates.

We were back on the same run, Bandar to Kharg Island, but on one occasion we went out of the Gulf with a cargo to Karachi in Pakistan.

Karachi used to be the capital of Pakistan, but when it became an Islamic State, Islamabad became the capital. It was also now an alcohol-free country, so I was warned by the Captain not to sell any alcohol to the locals.

After leaving Karachi, it was back up the Arabian Gulf to Bandar Mahshahr. I remember on one occasion quite a few of the hierarchy had come down to the ship from Tehran.

I was told by the Captain we would be entertaining them on board ship, with a meal provided for the guests and a couple staying on board in the stateroom to see the ship being loaded. Everything appeared to go well with the meal that was presented.

Not long after, the situation appeared to get worse with the Iraqis and Iranians. We were told by BP Head Office to move over to the Western side of the Gulf, and be ready to be repatriated if things got any worse.

Half of the Officers were put ashore in a hotel for a few days, whilst the remainder ran the ship. In the meantime, negotiations were going on to bring a full complement of officers and a full crew. I stayed on board as, if we did leave, I would have to close the victualing account plus the bond account, because these came under the agreement BP had with the National Iranian Tanker Company.

In the meantime, the Indian crew was repatriated back to India, and a couple of days later a Pakistan crew joined along with a full complement of German Officers, who would be taking over from the BP staff.

It all happened so quickly. One minute we were running a ship, next we were ashore waiting for a flight home. We signed off the ship on the 28th October, 1980.

After my next leave, I was to join the *British Poplar*, one of the tree boat class, in Ecuador.

On the M.T. *Tabriz* it was the only occasion where I had actually been around the world – we flew across to Alaska, down into Japan, sailed to the Gulf, then flew home to the U.K., therefore circulating the globe.

CHAPTER 16

To join the *British Poplar*, we flew from Gatwick Airport across the Atlantic to Venezuela, to Caracas, then on to Ecuador to the capital, Quito.

I was told beforehand that I was the only one joining the Br *Poplar*, but on the flight I noticed one or two people whom I knew. I got talking to them, but they were joining another BP Tanker at a different port in Ecuador. We were all staying at the same hotel in Quito, and two days later I was to join the Br *Poplar* in the Port of Balao. I was put in a taxi with the driver and his wife, or girlfriend, and we drove down to the coast. It took us about 10 hours to get down there, as halfway down we had to make a detour because the road was covered due to an avalanche.

We eventually arrived and I was put into a hotel for about three days, awaiting the arrival of the ship into port. It was a bit strange being on your own and not speaking the language. There really wasn't much there or anything to do. The local beach was dark in colour, more like volcanic ash than sand.

Eventually the ship came into port and the Catering Officer handed over to me and left to take his leave.

We left Ecuador after loading there, and headed towards the Panama Canal. After transit of the Canal, we were to go up the US Eastern Seaboard to New York.

Passage through the Panama Canal was entirely different from the Suez Canal. In the Suez Canal you normally steamed straight through it, but in the Panama Canal you went into locks, then into lakes. You might be in a lock with three or four ships at different levels; you could be looking down on them. It was unreal.

You would enter a lock and diesel locomotives on rails (known as mules) would secure the ship with wire ropes until you were in the next lock, and so on, until you entered the lake and sailed to the next set of locks. It was an ingenious invention, and so simple a way of transporting goods.

After transit through the Canal, we sailed up to New York. Once you hit the US coast, the radio stations were fantastic. Being an avid fan of Country and Western, I could have 24 hour, non-stop music. We went to one of the suburbs in New York, named Wilmington, and after dealing with the Port Authorities and ship's stores/provisions, we managed to get ashore. We visited one of the shopping malls there, which was huge. I had never experienced anything like it; you could have spent a week there and still not seen everything, or that was the impression.

After leaving New York, it was back round to Ecuador, via the Panama Canal. The thing I liked about the Panama Canal was that it wasn't as intense as going through the Suez Canal. There appeared to be pressure pestering to buy souvenirs from the boat men or bum boats in transiting the Suez Canal. Going through the Panama Canal was more simplistic.

After arriving at Ecuador, we went to a port called Guayaquil. I didn't know at the time that there was oil in Ecuador, but there must have been quite a lot for BP to get the contract for two ships to transport oil round to the US coast.

After loading the cargo at Guayaquil, we went back through the Panama Canal to discharge the cargo at New York.

On the way round I was informed that I and a few others would be relieved. It must have been the shortest voyage I had been on in the twenty years I had been at sea. I joined the Br *Poplar* on the 11th Jan and signed off on the 20th February, just over five weeks.

I was at home for a few days before being told that I would be joining a ship I didn't even know the company owned. They had bought a gas carrier, and diversified into Liquified Petroleum Gas (LPG).

We were to join the *Gas Enterprise*, as the ship was called, at Ras Al

Khaimah. It seemed strange joining this type of ship after having mainly been on oil tankers.

We loaded the ship in Saudi Arabia to take our cargo to North West Europe. The pipework on these gas carriers was colossal. The number of pipes that were fitted to get the cargo loaded looked like a maze, and the ship didn't appear to be loaded. After discharging the cargo, we went back to the Arabian Gulf to load the next cargo, and I was to be relieved. I was only on the *Gas Enterprise* for seven weeks, leaving at Ras Al Khaimah on the 25th April 1981.

After my leave, the next ship I was to join was the S.S. *British Dragoon* at Europort in the Netherlands.

Ships of larger tonnage could now transit the Suez Canal. The *British Dragoon*, being one of the regimental class and just short of 50,000 D.W. tons, was one of them. It was a big saving in time and cost going through the Suez Canal instead of travelling round the Cape of Good Hope.

After transit of the Suez Canal, the *Dragoon* sailed round to load cargo in the Arabian Gulf, at Kharg Island.

We then proceeded back to Rotterdam to discharge the cargo; it took about 21 days from the Gulf, via the Suez Canal, to the discharge port.

We did a couple of cargoes down to Nigeria, and back to the Isle of Grain. Then it was back through the Canal and round to the Gulf to load the next cargo at Mina Al Ahmadi in Kuwait.

After loading in Kuwait, we were to proceed back to North West Europe with the cargo to discharge at Hamburg in Germany.

After going through the Suez Canal and into the Mediterranean Sea, I was informed that I was to be relieved once the ship arrived at Hamburg. I suppose I was lucky on this occasion as it meant I would be home for Christmas. I handed over to my relief and left the ship on the 4th November, 1981.

After my leave from the Br *Dragoon*, my next ship was the M.V. *British Vine* – an Asian crew ship – and I was to join it in Swansea in Wales.

The ship was storing at the time, but luckily enough local riggers were doing the loading work. At the time the ship was mainly trading around the UK, and coasting round North West Europe. Almost every time you joined a ship, you would be tied down sorting invoices out from suppliers, starting new bond accounts, opening victualing records. You often found yourself stuck in your office for about a week sorting everything out, and bringing all the accounts up-to-date.

If you were on the UK coast you wouldn't be able to open the bond locker to issue a cigarette or tobacco ration unless the company paid the duty. And they would only do this so many days after the last one.

Once you left the UK coast you would be able to give a cigarette issue, as you were out of UK restrictions. The Serang, who was the head spokesman for the crew on an Asian crewed ship, would be up asking for a cigarette issue for the crew. The best thing about it was none of them smoked cigarettes; those that did smoke, smoked biddies. At our first port of call the cigarettes would be all flogged off to the highest bidder, without the local Customs knowing.

You had to be careful what you were doing as you could be in port every two to three days. So I stuck to the company's restrictions and only allowed them 200 cigarettes every 10 days, otherwise we would have run out, because about 35 crew members came up every time there was an issue. Trading around North West Europe, the Scandinavian countries and up the Baltic Sea could be quite severe in winter months. Sometimes it was so bad that the seas would freeze over and you would need ice breakers to get you in and out of the Baltic ports. On a few occasions an ice breaker would smash the frozen sea and take a convoy out of the Baltic Sea.

When the weather picked up around the Scandinavian countries, and you were in port for a couple of days, you might be able to arrange a football match with another foreign ship. The Scandinavian countries appeared to have a very good seamen's mission organisation. One particular time in Sweden they organised a sports day and invited numerous visiting ships to take part.

Sometimes on Asian crew ships they could repatriate the crew home. Normally they would be on the ship for about a nine month contract, before they would be relieved, and a new crew would join the ship. The new crew would join with practically just basic essentials, whereas the crew leaving would be loaded up with watches, radios, cassettes and other electrical goods.

They would normally be repatriated from the UK, with about three or four buses taking them to Heathrow, unless it was more convenient to pay them off on the Sub-Continent.

The crew joining would have to be fitted out with new linen and bedding, and – especially up the Baltic – with heavy weather clothing for the deck hands, and new boots for all the crew. It was quite an ordeal getting them all fitted out. The crew joining would be there about two months before they could afford to send a remittance home to their families.

After being on the *British Vine* for just short of three months, I was relieved of my duties and was to leave the ship at Europort in the Netherlands on the 5th April, 1982.

While I was home on leave, the Argentine Government was doing a bit of sabre-rattling about the British Government handing back the Falklands to Argentina. So many of the UK Merchant Fleet were seconded to the Royal Navy that BP had lots of their smaller tankers being used to transport certain types of fuel to the naval ships being used down in the Falklands, as well as other tankers and cargo company ships.

CHAPTER 17

The next ship I was to join was the *British Hawthorn*, another one of the company's tree boat class.

I was to relieve the permanent Cat/Officer while he went on leave. I joined the Asian-crewed ship at Europort, where I had left my last voyage on the Br *Vine*.

The Br *Hawthorn* was also coastal trading around the UK and North West Europe.

On one occasion the ship took cargo to Grangemouth in Scotland, and we were storing the ship while we were there. Sometimes representatives from breweries visited the ships. BP used to purchase Tennents lager and three other lagers for the bars on their ships. On this particular occasion, a rep visited from Tennents Brewery and asked me if I had any free samples sent home, which I hadn't. The next letter I had from my parents said that 24 cans of lager had been sent home; they went down a treat next time I was on leave.

We had a few cargoes round North West Europe, mainly Antwerp, Rotterdam, Hamburg, Gothenburg and Copenhagen. The good thing about coasting was that the time went quickly. I was told I was leaving the *Hawthorn* when the ship docked back at Grangemouth, but I was to join the M.V. *British Fern* in Milford Haven in Wales.

I left the *Hawthorn* on the 10th September, 1982, and I was to join the Br *Fern* on the 15th September, so I managed to get three days at home on the way down from Scotland to Wales.

Joining the ship in Wales was a long slog. The taxi wasn't allowed onto the jetty, so you had to hump your case down to the ship. And, after closing the accounts on the *Hawthorn*, I was to open the accounts on the *Fern*.

The trouble with Asian crew ships was you had to have your wits about you, or you could easily be taken for a ride.

For instance, when opening the bond and victualing accounts, you could be tied up in your office for four or five days and become unfocused when issuing stores. With your head in paperwork, all you wanted to do was give the stores out and get back to your office.

The Scullion, whose job was the least demeaning in the Catering Department, used to collect the stores. For about four consecutive days he asked for tea bags, with about 250 tea bags in a box. I eventually asked him what he was doing with all these tea bags, as there were only about 15 Officers on board ship, and the Catering Dept and the crew were issued loose tea. I put a stop to boxes being issued and instead gave him loose tea bags. I also found out later on, when I was walking down the jetty joining the *Fern*, that a Deck Cadet had seen one of the Catering Staff selling a jar of coffee to a jetty worker.

There were signs up all around the ship's accommodation that stores were for consumption on board ship only. The trouble with this Scullion was that all the other staff in the Catering Dept were unsure of him, as he was well educated in their caste society in India. The impression you got when you issued stores was that if they weren't used, the goods were theirs.

Sometimes on Asian crew ships, the amount of rice issued would accumulate where they kept it on board ship, and they would take extra M.K. mutton in lieu of rice, which the company would allow.

Being in and out of port every two to three days in different countries, you would have to fill Personal Manifest Declaration Forms in for everyone on board at each port.

You personally had to get the crew organised, and fill in their Declaration of Items on the manifest – watches, radios, cassettes, etc – and this could take three or four hours. Some couldn't sign their names so you would have to use their thumbprint. When it was time for leaving the ship, you would have to do a staff confidential with the head of each department, Catering, Deck & Engine Depts. This was sent off with the crew when they were repatriated back to India.

The Scullion who had caused a few problems while I was on the Br *Fern*, wasn't too pleased when I read him his Review Form. He asked me to change the report or he wouldn't sign it. I told him whether he signed it or not, it was still being sent in.

It was getting near my time for being relieved. We had loaded a cargo in Germany, but hadn't had any orders where to discharge it; some speculator had bought this cargo and put it on the market to be purchased, in the hope of making a profit.

Eventually we discharged the cargo at Flushing and I had about 24 hours to close the bond account and the victualing account, as I was being relieved. I left the *British Fern* on the 19th December, 1982, delighted that I was not going to miss Christmas at home.

On this occasion on leave, the company decided to send me and a few others from different companies on a Fire Fighting Course on the Tyne at South Shields.

It was similar in structure to the one I had attended in Plymouth in 1971. This course actually had a steel frame in the shape of a ship's accommodation, with smoke-filled spaces and a more realistic atmosphere about it.

Both courses gave me a frightening experience of dealing with fighting fires.

During the late 70s, shipping companies had built VLCCs. But by the 80s there was a slump in the amount of oil needed, so shipping companies started laying ships up, including BP.

The next ship I was to join was in Brunei. BP had diversified into the lay-up service.

I flew to Brunei on the 28th March to join S.S. *British Purpose*, at Labuan. BP had two VLCCs tied up alongside each other, fore to stern, with BP manpower. When staff joined laying-up ships in Brunei, they were put on a boat with their luggage, and left Bandar Seri Begawan for the port of Labuan.

Out in Labuan, there were two Catering Officers – one on each ship – but all the food was prepared on the Br *Purpose* with a full Indian crew.

The company would notify the Operation Management Team in Labuan that a ship would be coming up to the lay-up unit. The ship would then be shut down by an engineering team and de-humidifiers would be put on board.

The ship in the lay-up would be routinely checked weekly, monthly or quarterly, and monitored regularly. All the air would be taken out of the vessel and everything sealed up like doors, vents, etc.

The majority of the ships came from Singapore. Occasionally they would bring stores up for the lay-up service, being a more competitive price in Singapore. Also they would bring bonded stores, lagers, cigarettes, spirits for the bars, maybe about 2,000 cases of lager, which would be transferred to the Br *Purpose*. Any stores remaining on the ships coming to the lay-up base would also be transferred across, with a fee agreed for the cost.

The base where the ships were laid-up was ideal, in a bay. Labuan was the nearest place to get ashore and BP had about four crafts for getting about in the bay, and to get to the laid-up ships, which varied in size.

Once a week I had to go ashore to get the weekly vegetables and salads. The first time I was taken to different shops and agents by the other Cat/Officer, to show me the ropes.

Mail was also sent ashore once a week and films were changed. If someone had taken a few photographs, I would put the film into a photograph shop, and the photos would be ready in less than an hour.

Once you got into the swing of everything, it was quite a good set-up there.

Once a week you could go 'on the jolly', as it was called; a new name for a piss-up. Depending on how many went, the boat would be loaded up with barrels filled with ice and beers and set off to this island. We'd anchor the boat up, get all the gear off, get the barbeque set up and have a swim or a game of cricket. The weather was great, but you knew about it the next day as it felt that your face had been shrunk-wrapped.

Every so often we would have a game of football at Labuan against one of the local teams. There were quite a few ex-pats out there from the UK, and Aussies. They had their own club and you would be invited in on occasions.

Every couple of months there would be a barbeque on the ship. That sounds unreal having a barbeque on an oil tanker. But they were quite well run and the locals were ferried from ashore out to the ship and back. The highlight of the year in Labuan was the annual hash run, as it was called. We know them as marathons.

It covered about 10 miles on one of the larger islands and was well organised. It would cost about £5 entrance fee and, if you finished it, you got a t-shirt. There was also a barbeque and a couple of drinks afterwards. Some of the locals and ex-pats took it quite seriously, as the winner was presented with a trophy.

After being there for a couple of months, we went on the weekly jolly. As it wasn't very large, I decided to walk round the island. You could practically walk all the way round, apart from a small stretch, where you had to swim to the next part of the beach. Unfortunately I went on my own.

When I had got about halfway round, I was walking along the beach when up ahead, coming out of the sea water, were these manatees/lizards. They must have been 8-10 feet in length. When I saw them, I was having kittens; I hoped they weren't carnivorous or I was dead.

Luckily, they went into the undergrowth and ate the vegetation. If I had been in the marathon that day, I would have won it. I was relieved to get safely back to where we were having the barbeque, but it was an experience I will never forget.

Near the end of my tour of duty in Labuan, BP brought two or more of their VLCCs to lie-up. That made four tankers in tonnage, and 800,000 tons of shipping laid-up in the anchorage which BP was looking after; between 40 to 50 ships of different size.

Near the end of August I was notified that I would be relieved of my duties. I signed off the ship with a few more Officers who were going home on leave, on the 27th August, 1983.

CHAPTER 18

For some unknown reason, this time on leave I was at home longer than normal – nearly four months. But eventually I was told to join the M.V. *British Test* at Croydon, on 23rd December, 1983. This was one Christmas I was definitely going to miss at home, but I really couldn't grumble as I'd had the last three on leave.

The *British Test* at the time was coastal trading around Europe for a couple of weeks. We then loaded a cargo from the Isle of Grain to take to Sicily in the Mediterranean Sea.

After discharging the cargo, we got orders to load a cargo in the port of Porto Foxi in Sardinia, and take it across the Atlantic to Boston in the USA.

After leaving the Mediterranean, it was about ten days across the Atlantic. It felt good to get back to the routine of being at sea for a spell, having been on a ship at anchor for four months, and the previous three ships mainly coasting around North West Europe.

On arriving at Boston, the ship was invaded by the Port Authorities, Customs and Port Health. We should have been at Boston for about 36 to 48 hours and discharged the majority of the cargo, and then the remainder of the cargo would be taken down the coast to Portland Maine.

One of the Port Health Restrictions, when ships enter the USA, is that all rubbish bins must be inside the ship's railings, and the bins covered. Prior to discharging the cargo, samples were taken and it was found that the cargo was contaminated with water.

So some of the cargo was pumped ashore and then decanted to get rid of the water.

This meant that instead of being in Boston for 48 hours, it was anybody's guess how long we would be there. After about four days the rubbish was starting to stink, as there were no facilities in the Port of Boston to take it away.

The Port Health also had to come down to the ship to release the food from the freezers on board, as these had been sealed up on entering the port.

I was dousing the bins down with disinfectant, trying to keep the smells under control, but maggots were climbing up the outside of the accommodation.

A couple of days after the maggots, the ship was abuzz with bluebottle flies. Eventually – after eight days – we left Boston and went to Maine, which was about 24 hours away. The only problem was we weren't allowed to dump the rubbish that had accumulated until we were 24 hours out of US territorial waters.

Luckily we were only at Portland Maine for about 24 hours, and everybody was glad when we left the US coast to get rid of the rubbish and wash the ship down.

From there we travelled to Venezuela to load a cargo to take to Rotterdam. On the way across the Atlantic, I was informed that I would be relieved when we arrived there. I didn't realise at the time that this would be my last BP Tanker, M.V. *British Test*.

On arrival at Rotterdam, I handed over to my relief, and signed off the ship on the 26th March, 1984. When I got home, my father had been taken ill. They didn't say at the time that it was cancer, although he was going for treatment in Leeds.

I wasn't surprised. He'd started work at the age of 14 down the mines, and retired at 64 years. 50 years in the coal mining industry was enough to kill any man off, in the conditions they had to work in. I had taken over running his garden over the last couple of years, as he was getting on in life and found it was too much work at his age. I still dig the garden and grow vegetables and bedding plants; I would sooner see a garden tidy than untidy. As these houses were built just after the First World

War, the gardens were quite large to ensure that people could grow their own vegetables. Unfortunately, times have changed, and I think I am the only one who now grows vegetables in St Helen's Avenue.

After being at home for three months, I was notified to join a ship called the M.V. *Vic Bilh.* We flew out to Houston in Texas, and it was practically a full crew change, with three or four Officers and a couple of wives.

At Houston, we cleared Customs and had to take a bus down the coast to Port Arthur, where the ship was to berth. I didn't realise at the time, but the crew who were joining the ship were permanent on the *Vic Bilh* and were rejoining after being on leave.

The Catering Officer was the same one I had relieved on the M.V. *Shirvan* while he went on leave, Dave Wooley. I mentioned the system he had introduced on the Shirvan, which BP Storing elaborated on, and that he never got any credit for. He said I was the only one who had ever given him any credit for doing so.

Just as we pulled up on the jetty, the ship was fully alongside and the gangway down, so it was good timing. The Catering Officer I was relieving was going home on leave and I would be taking over for two years.

Once on board, I got into my cabin to get sorted out, and not long after the pipe lines were attached in preparation for discharging. On a previous occasion a ship had gone down the river from Port Arthur, and the *Vic Bilh* had started surging up and down, due to the other ship's speed going down the river.

Apparently the Captain, who was a Scot, had phoned the Pilot Station and asked if any more vessels came down the river could they reduce speed, as the previous ship had caused the *Vic Bilh* to start surging and we were about to start discharging the cargo.

I will never forget what happened next. I was looking midships and the ship started surging; we had started discharging the cargo. Next thing, one of the shore pipelines became disengaged from the ship's connection and crude oil was spewing into the river and onto the jetty.

Then the electrical stanchion (on the jetty) somehow was dragged down onto the jetty, and it was as if someone had thrown a jumping cracker amongst the spilt crude oil and it went up in flames. I was watching all this unfolding in front of my eyes as I looked out of my cabin.

Of course, all hell broke loose. Luckily, the crew going home had just left the ship and were on the wharf boarding the bus to take them up to Houston.

All down the left hand side of the *Vic Bilh* was ablaze; down the same side were about 10 Yokohama fenders in rubber. These all caught fire and made the situation ten times worse, with smoke billowing up into the atmosphere. By this time, the pumps on board ship had been shut down, so no more crude oil was being pumped ashore. The oil that had gone into the river had also caught fire and it appeared that all the left hand side of the ship was ablaze.

Thank God the crew who had rejoined the ship knew where everything was stowed and were familiar with the equipment they were using. I must give praise to Dave Wooley for the professional way he handled the form cannon, helping to put the fire out.

During the fire on board, all the staff that had not been on the *Vic Bilh* before –including the wives – were informed by the Captain to assemble on the starboard side of the ship and to disembark by lifeboat. While all this was going on, not one attempt was made by the Port Authorities to help the ship's personnel to put the fire out. A tug appeared but stayed well out of reach in the middle of the river.

When all the staff were in the lifeboat, we were lowered down to river level ready to make haste away from the *Vic Bilh*. Unfortunately, one of the ropes got wrapped round the lifeboat's propeller and we couldn't move.

By this time the fire down the port side was astern and appeared to be coming towards us in the lifeboat. Fortunately for us, there was a small boat in the river that came to our assistance and towed us further up the river to safety.

After about two hours they let us back on board the *Vic Bilh*, the fire having been put out entirely by the professionalism of the Captain and the crew. I often wondered how the refinery didn't have any damage at all inflicted on the structure. But I couldn't believe the damage to the ship. Climbing back up the gangway, it was a mess; all the fenders from the forward to aft were just cinders.

Up in the smoke room, one of the Refinery Managers, I assumed, came in and said to the Captain, 'You've done one helluva job today, probably saved Port Arthur from blowing up.' The refinery was colossal, as we found out a few days later as we went through it.

We were then taken off the berth and put on another berth further down the river to assess the damage.

The M.V. *Vic Bilh* was owned by an American Company, registered in London to Fairfield Maxwell Services Ltd, on charter by Elf Oil Company of France, and manned by BP staff.

It was net registered tons 41,800 and able to go alongside other ships and offload about 70,000 tons, therefore allowing VLCCs to get into ports they normally wouldn't manage. One of the main characteristics of the ship – the Yokohama fenders down the port side – was on davits. These fenders would enable the ship to go alongside another ship and use the fenders as a buffer while offloading so much of the cargo.

These fenders on the *Vic Bilh* were in two sizes; the largest of the two came at a price of around £25,000 each.

After the ship was assessed for damage and could still pump the remaining crude oil ashore, we went back up to the refinery to discharge the remainder of the cargo. The crude oil that had been lost at the start of the incident before shutting the pumps down was negligible, but the damage to the ship looked horrendous. All the paint down the port side of the ship had been burnt off and it was bare steel that was showing.

Eventually the cargo was discharged at Port Arthur and we left the US to go to the Gulf of Mexico. It was ironic that we went to the same area where BP Oil Company experienced one of their worst oil spills on record 25 years later.

It's a pity the US Government didn't come to the assistance of the

Vic Bilh when the fire broke out at Port Arthur.

While in the Gulf of Mexico, the loading of the *Vic Bilh* was by submarine pipeline. You tied up to buoys and then connected the pipes, and the oil was pumped into the ship. While there, I received mail from home saying my father's condition had got worse and he was having chemotherapy for terminal cancer.

Thankfully, my mother and sister were at home to take care of him. After leaving the Gulf of Mexico, we went across the Atlantic and back to Europe to a port in France, Donges. In the meantime, the company which owned M.V. *Vic Bilh* was making enquiries about putting the ship into dry dock after the incident at Port Arthur.

When we arrived at Donges, you would have thought the ship had the plague by the reaction of the locals. I must admit, as I hadn't seen the port side of it for about three or four weeks, it did look somewhat the worse for wear.

After leaving and discharging a couple of cargoes in the Mediterranean, we were to discharge a cargo at the port of Fos, in Southern France. There was also talk that, after discharging the cargo, the tanks would be cleared because a dry dock had been secured for the ship to go into at Palermo in Sicily. First we had to berth the ship at Fos, then deal with the Customs, Ship Chandler and Port Authorities, and ship's staff visiting the doctors, etc.

I got the drastic news that my father had passed away on the 4th September, 1984, when a couple of personnel staff came to speak to me. I was to be sent home on compassionate leave.

After sorting the accounts out on board ship, I was to leave the ship on the 5th September; at least I would be at home for the funeral on this occasion. Back home, there wasn't much to do as the funeral arrangements had already been taken care of. There was a delay in the cause of death on the death certificate, which my brother had taken care of at the Coroner's Office. The actual cause of death on the certificate was put down to Bronchopneumonia. After the church service, my father was buried in the same grave as my youngest brother.

A reception was held at the West End Club, where my father had been a member all his life and on the committee for many years.

The house seemed sombre. When you lose a member of the family, it takes a while to get used to. I threw myself into the garden, to get it sorted out before I went back to sea.

CHAPTER 19

Eventually, I was notified to rejoin the *Vic Bilh* at Palermo in Sicily around the middle of October. Thankfully my sister was at home with my mother for support while I was away.

I re-signed on the *Vic Bilh* on the 18th October, 1984. Although the ship was owned by Fairfield Maxwell Services Ltd, BP managed and did the catering, and ran the ship.

On this occasion, all the food and bonded stores were sent out by refrigerated lorries from the UK, apart from locally sourced vegetables and salads. It must have been cheaper and less costly to send out from the UK than to purchase locally.

There was a vast improvement to the damage on the *Vic Bilh* by the time I arrived; all the Yokohama fenders had been replaced. I don't know if it was true but the total damage from the fire at Port Arthur was estimated to be in the region of seven million US dollars.

After leaving Sicily, we went to Nigeria to load a cargo to bring back to France. During the voyage, BP had sent out a time and motion study for all the Officers and Senior Crew Members for a period of six weeks, to justify the hours that each one was working every day, seven days a week. It was different being on a managed ship to a BP Tanker; particularly the amount of storage space on a BP ship and the amount of spare parts.

The one good thing I got out of the whole time and motion study was the Captain complimenting me on making a true statement of the hours I worked throughout the six weeks. It certainly justified my position/job at sea, if that is what the whole procedure was about.

During the voyage the Senior Management on board ship – Captain, Chief & 2nd Eng, Chief Officer, Radio/Officer and myself – used to have a weekly meeting, as on BP ships, to discuss the work routines for the week.

When the Captain asked me where I was from, I had told him Hemsworth, and he said that he'd heard there had been some trouble at one of the local pubs in Hemsworth, The Blue Bell Hotel. It was at the time when the mine workers were out on strike, and the police had apparently marched into the pub as a show of strength to sort a few hotheads out, but it had backfired on them. The Captain mentioned that it had been on the BBC World Service on the Sunday morning.

During the duration of the miners' strike, I sent a donation to the landlord of the Blue Bell at the time, John Smallman, to buy a drink for the regulars, as I used to frequent the Blue Bell for a pint of Tetley's when I was at home.

With the ship back up to working standards, I saw the full system of how the Yokohama fenders worked when we went alongside a VLCC and offloaded 70,000 tons at Lyme Bay, just off the coast of Lyme Regis. This place was a good anchorage to do this type of procedure, and we then went round to Milford Haven to discharge the cargo. After leaving Wales, we were to go to Gabon on the west coast of Africa; this was one of France's old colonial outposts in Africa.

The Captain who had been on the *Vic Bilh* when we joined at Port Arthur, had left. The permanent Captain, who had joined with me and the permanent crew, had gone on leave and was replaced by the Captain who, in my eyes, had saved the ship from destruction at Port Arthur. His name was KGE Lawrence, a Scot. There was talk that a reimbursement would be paid to Officers and crew members pro-rata from the top downwards, for the gallant way they had fought the fire on the *Vic Bilh*.

After a few ports around the Mediterranean, we loaded a cargo to be discharged at Europort in Holland. I was told that I would be relieved of my duties when we arrived there, and go home on leave. While I had been away at sea this time, my mother and my eldest sister had gone for a holiday to the Canary Islands. Unfortunately my mother had had her handbag snatched; it was the second time my mother had been mugged, once before in Hemsworth and now abroad.

I eventually left the *Vic Bilh* at Europort on the 22nd February, 1985. Back home, I took a couple of weeks to adjust to life then got stuck into the garden, sorting things out and getting it ready for the growing season. Unfortunately all good things come to an end, and I was to rejoin the *Vic Bilh* in Texas.

We flew there at the back end of June. I had done well with my leave on this occasion – just over four months at home, so I wasn't complaining. The port in Texas where the *Vic Bilh* was discharging her cargo, was aptly named Texas City. This was another place in Texas which suffered a really bad fire at the refinery a few years later on.

After leaving Texas City Oil Terminal, we went to Venezuela to load a cargo to take across the Atlantic to an unnamed port in France.

The thing about shipping companies was that it wasn't practical to have a ship empty, so you would go across the Atlantic fully laden, and the same coming back. After discharging the cargo at Port de Bouc in France, we were to go to Nigeria to load the next cargo.

To pass the time at sea and if the weather was good, particularly on the VLCCs, we would have a game of cricket on the ship's main deck. We used to have a game on the *Vic Bilh*. We made the bats and stumps, and the balls would be made from rope and shaped. It got to be the highlight of the week, with a buffet after the match.

On the *Vic Bilh*, there were rumours that the BP Shipping Company were in talks with various other agencies about putting their fleet of tankers out on what was known as Agency Manning.

BP Shipping Company was one of the group's organisations which wasn't making money out of its fleet of oil tankers. It was being run at a loss, whereas the oil exploration and the refinery's capacity were where the company was making the vast profits.

The permanent Captain had rejoined the *Vic Bilh* in between ports, and we appeared to be doing most of our trading between Nigeria and

French ports, with the ship on charter to Elf Oil Company.

I was to be relieved of my duties on the *Vic Bilh* at the port of Donges in the South of France in December, 1985. I had been on the ship just short of four months. The time away from home appeared to be getting shorter, and the time home on leave seemed to be longer. I didn't realise at the time, but this was to be my penultimate voyage at sea.

For some unknown reason I managed to be home for Christmas, whether it was down to the *Vic Bilh* trading pattern, I just didn't know.

I was notified that I was to rejoin the *Vic Bilh* at Teesport, near Middlesbrough, about the 6th February, 1986.

After leaving Teesport, we sailed to Gabon to load a cargo. Gabon wasn't a large oil producing country, after the likes of Nigeria.

The rumours relating to BP Shipping putting their oil tanker fleet out to agency manning, was turning into reality as the voyage progressed. It meant for the Officers that the fleet would be split up with so many ships under different agencies. The terms and conditions would be renegotiated, and there would be an increase in wages annually. Voyages would increase to five or six monthly tours of duty, fellow Officers would be of different nationalities. Over the next few weeks, morale from top to bottom of the ship's personnel took a nosedive, as everybody felt in limbo about what was happening.

It felt as though the company had neglected the ship's personnel. Negotiations were going on with BP management ashore to transfer the shipping over to the different agencies taking over and, until everything was sorted out, we onboard ship would be the last to know what was happening. While all this was going on, the ship was still functioning and being run as before, although there was a lot of uncertainty.

Eventually all the ship's staff were asked if they would be re-signing under the new terms negotiated. I personally would not be re-signing, as the terms and conditions were not to my liking. So I would be taking the redundancy terms that were offered, and would be terminating my contract with BP Shipping.

If I had re-signed under the new terms and agreement, I would have been demoted to Cook/Steward, and ended up doing the accounts for the catering stores as well as cooking for the ship's staff.

The extra work would not have bothered me, but I felt then that it was the right time to finish with the Merchant Navy, and this was the ideal opportunity to do so.

I also thought that I would have no problem getting established in the job market with the qualifications I had.

Eventually we were told the place and port where we would be leaving. We would be discharging the cargo at Bilbao, and the staff who were not re-signing under the new terms and agreement would also be signed off.

Before leaving Bilbao, sorting the Customs, Ship Chandler and other Port Authorities out, I issued all the new crew with laundry and safety boots and overalls. I sorted the new cook out, showed him as much as I could relating to his job, and left the *Vic Bilh* just before the ship left Bilbao, on 3rd April 1986. Twenty-five years – a lifetime at sea.

After a couple of weeks at home to unwind, I went down to the Jobcentre to sign on. I wouldn't be due any unemployment benefit for 12 months with the redundancy payment, but I was due my National Insurance Stamp.

I couldn't have picked a worse time to look for a job ashore in the Catering Trade. Not long after I finished at sea, there was an outbreak of salmonella food poisoning at a couple of the local hospitals. A big clampdown was ordered. All staff working in catering establishments run by the local authorities/councils had to have 706/1 Part 1 and 706/2 Part 2 plus a certificate in food hygiene, to work in schools and hospitals, etc.

If I applied for one job, I must have applied for over one hundred in the next 12 months. I either did not have the proper certificates, was over qualified, too experienced, or did not have the right qualifications. I did eventually get a few part-time jobs.

I went to Wakefield College and achieved the qualifications in City & Guilds 706/2, and the Food Hygiene Certificate. One of the college tutors said he didn't know what I was doing at the college, as I was more qualified than he was. The problem was, I didn't have the certificates to say so.

SUMMARY

After being at home for a couple of months and no prospect of getting any employment, there was only so much to do. Most of my time was spent in the garden or going for long walks in the surrounding countryside. Locally, down at Vale Head Park, the local authorities in Wakefield had decided to create a small, 9 hole golf course, which I used to use frequently and still do.

The local park at Hemsworth was well used by the locals, and all the parishioners of Hemsworth, Kinsley and Fitzwilliam and surrounding areas.

Hemsworth Vale Head Park was opened in 1932. And an extension adjoining Vale Head Park was opened in the mid-1980s, known locally as "The Water Park" and opened by Prince Charles.

Around the middle of 1986 I had applied to Wakefield College to enrol as a student on the City and Guild 706/1 & 2 course, on a day release for two years. During the time I was on the course I was approached by a local nursing home for the elderly and offered part time employment as a cook. I accepted the job on condition that I was released for one day a week to attend the college during the course.

I started work on a Sunday morning, but the only problem was that I didn't get paid for 5 weeks. The job itself was quite basic – weekend on, weekend off – but at least I was employed. This particular job lasted about eight months after an incident that happened, when I was paid eight hours short in my monthly wages. I notified the Sister, as you practically never saw the owners, and she said I would receive it in my next month's salary, which I accepted.

I did notify the Sister on a few occasions to let the owners know that I would be leaving if I didn't get what was due to me. When I didn't receive my eight hours' pay, I left. The owners phoned my house on three separate occasions making excuses, when it was their accountant who made the error due to a bank holiday which should have been double time. The case went to tribunal, but I lost the case, plus my employment benefit which I was now claiming. For months on end the job prospects were a bit grim, with over 3 million unemployed – similar to the present climate. I managed to get a part time job on a local scheme which had been set up with the District Council, Business Commerce, the Church, etc. It was a base in Wakefield where management, trade personnel (bricklayers, plasterers, labourers, etc.), were trained and put to use in the community.

Gardening schemes were set up, and I was put in the stores with a supervisor. Sometimes the supervisor and I were sent out to pick up furniture that had been donated and distributed to hard-up families. This short-term labouring job lasted about six months, consisting of about 20 hours per week, but with the day release at college plus the gardening, it kept me occupied.

One of my biggest mistakes when leaving the Merchant Navy was not taking up driving lessons. Going to Wakefield College on the bus was a nightmare when the school run was on. Normally it would take 25 minutes, but during the school term, sometimes it would be over an hour! During the two years attending Wakefield College, the majority of people on the course were from schools and hospitals.

After passing the City and Guilds and Food Hygiene Certificate in 1989, I was finally qualified to work in the food industry ashore. For some unknown reason, I fell into another part time job, this time at a pub/restaurant.

The pub was in a very good location with a fishing lake and lawned garden area. It should have been an ideal spot for weddings and lawn parties, etc. The only problem was the lack of transport for myself to get home after 10.30pm.

Sometimes communications were bad. One Tuesday evening I arrived to start work. Nothing was in the book relating to any bookings that evening. The waitress arrived an hour or so after me and told me a busload was arriving with about 50 people coming from York Races. You can imagine what frame of mind I was in when I heard. As it happened, we managed to get them all fed and watered.

Another time a wedding was taking place there. Being a buffet lunch, the bride and groom and parents were surveying the layout, when the owner's wife said to blame me if there wasn't enough food to go round. You can imagine what thoughts were going through my head.

Not long after working there, the local greyhound stadium was having a restaurant built adjoining the stadium, in the village of Kinsley. I don't know what it was about Kinsley, but in my eyes it had the best entertainment for miles around. You could go down there every night and something would be going on, be it The Farmers, the Working Man's Club, The Willow Club or the Coronation Club (known as the "Ping Pong"). There was also greyhound racing two or three times a week.

At The Willow Club, good quality entertainers used to appear, including Gerry Dorsey (better known as Engelbert Humperdinck), Ronnie Dukes and Ricki Lee (a very popular duo). Unfortunately, The Willow Club and The Farmers Club closed.

Whilst still working at the pub, I bought a small scooter for getting backwards and forwards to work because a few times the late bus didn't turn up or I missed it by working late, and faced a 3-4 mile walk home.

I stayed at the pub for about nine months. I was being encouraged to go down to the greyhound stadium to apply for a job in the restaurant which was nearing completion. After being interviewed, I was offered the post and gave my notice at the pub. The new job was only for three days a week, but that suited me as my mother was getting on in age.

The first impression I had when seeing the set-up in the restaurant was that it looked similar to Walthamstow Greyhound Racing Stadium, but on a smaller scale. The restaurant was laid out on two levels – a cafeteria-style menu, self-service, hamburgers, curry and rice, chicken and chips, etc; and the restaurant on a lower level, with an à la carte menu and a good view of the greyhound racing.

It took a while to take off but once the word got round, it got pretty hectic some nights. I personally thought the kitchen wasn't large enough to cope with the number of clientele that had been booked in; the clients could vary from 60 to 120, even more when functions were organised.

In the 1960s through to the 1980s, the governing body for NGR (National Greyhound Racing) used to, and still do, have the annual greyhound derby – the top prize money being £10,000.The owners of the Kinsley Greyhound Stadium decided to run a race with a total prize money of £20,000. At the time the stadium was classed as a 'flapping track' – it wasn't run on the same rules as the NGR, so they weren't governed by their rules.

The race was set up with all the trials, heats and number of dogs entered. Greyhounds were being sent from all over the UK. You can imagine the enthusiasm being generated with double the prize money for the winner. On a 'flapping track', it took about three months to set up with all the heats through to the final. On the night of the final, it had been decided by the owners that a buffet would be put out for the paying guests in The Jubilee Restaurant.

The money generated would pay the prize money for the winning owners. It was so successful that they decided to have another Kinsley Derby the following year.

I remember coming home one evening from the greyhound stadium and I was pulled over by the police to be breathalysed. As it happened I was okay, but when I arrived home, I noticed that my mother's handbag strap was snapped. It wasn't until the next morning that I found out that she had been mugged – that was the third time, twice in Hemsworth and once abroad.

As time progressed, my mother's health started to deteriorate. Fortunately for me, my elder sister was living at home, so between us the burden wasn't so heavy. Also, my younger sister lived close by and was a big help.

I worked at the greyhound stadium for about five years in total. Eventually it became part of the NGR. As time passed, my mother got worse and eventually we brought her bed down into the dining room for more comfort.

One night, she was rushed into the local hospital at Pontefract and admitted for tests which confirmed she had had a heart attack. After recovering, she was sent home and then re-admitted for further tests. By this time I had finished the job at the stadium. I hadn't realised how demanding caring for the elderly was, but luckily there was family at home to take care of her.

My mother was then admitted into the local hospital at Southmoor in Hemsworth. Fortunately you could visit there when you liked. Over the coming weeks she deteriorated – it was as if she was saying that she had had enough. On the 25th May, 1995, my mother passed away peacefully, with the majority of the family with her, in Southmoor Road Hospital.

After the funeral it was hard to understand how to cope with the void that follows after somebody close passes away – one minute they're there, the next minute they're gone. Luckily, I had plenty to do in the garden to keep me occupied. I didn't take any more jobs. I paid my Voluntary National Insurance Contributions until I was 60 years old to qualify for my old age pension when I reached 65.